HARCOURT

Math

Family Involvement Activities

Grade 6

Orlando • Boston • Dallas • Chicago • San Diego
www.harcourtschool.com

Printed in the United States of America

ISBN 0-15-320851-1

1 2 3 4 5 6 7 8 9 10 082 2004 2003 2002 2001

CONTENTS

UNIT 1: NUMBER SENSE AND OPERATIONS

Chapter 1: Whole Number Applications FA1
Chapter 2: Operation Sense FA5
Chapter 3: Decimal Concepts FA9
Chapter 4: Decimal Operations FA13

UNIT 2: STATISTICS AND GRAPHING

Chapter 5: Collect and Organize Data FA17
Chapter 6: Graph Data FA21

UNIT 3: FRACTION CONCEPTS AND OPERATIONS

Chapter 7: Number Theory FA25
Chapter 8: Fraction Concepts FA29
Chapter 9: Add and Subtract Fractions and Mixed Numbers FA33
Chapter 10: Multiply and Divide Fractions and Mixed Numbers FA37

UNIT 4: ALGEBRA: INTEGERS

Chapter 11: Number Relationships FA41
Chapter 12: Operations with Integers FA45

UNIT 5: ALGEBRA: EXPRESSIONS AND EQUATIONS

Chapter 13: Expressions FA51
Chapter 14: Addition and Subtraction Equations FA55
Chapter 15: Multiplication and Division Equations FA59

UNIT 6: GEOMETRY AND SPATIAL REASONING

Chapter 16: Geometric Figures FA63
Chapter 17: Plane Figures FA67
Chapter 18: Solid Figures FA71
Chapter 19: Congruence and Similarity FA75

UNIT 7: RATIO, PROPORTION, PERCENT, AND PROBABILITY

Chapter 20: Ratio and Proportion FA79
Chapter 21: Percent and Change FA83
Chapter 22: Probability of Simple Events FA87
Chapter 23: Probability of Compound Events FA91

UNIT 8: MEASUREMENT

Chapter 24: Units of Measure **FA95**
Chapter 25: Length and Perimeter **FA99**
Chapter 26: Area **FA103**
Chapter 27: Volume **FA107**

UNIT 9: ALGEBRA: PATTERNS AND RELATIONSHIPS

Chapter 28: Patterns **FA111**
Chapter 29: Geometry and Motion **FA115**
Chapter 30: Graph Relationships **FA119**

Using the Family Involvement Activities

The *Family Involvement Activities* provide you with a means of communicating with family members concerning what their children are learning in mathematics. In addition, these materials provide models that enable parents and other family members to assist students in the learning process.

The *Family Involvement Activities* serve several purposes:

- to provide instructional materials that enable older sibling, parental, or other adult tutoring of new concepts introduced in the student text;
- to provide a home math activity that students can complete with family members;
- to provide practice of new concepts in the form of homework;
- to provide a concept-based math game for family members to play together.

It is helpful when parents understand their role in their child's success in learning mathematics. To facilitate parents' understanding of what students are learning, an initial family letter is provided, which is accompanied by an important reference parents may use throughout the school year: the glossary from the student textbook.

In addition to the letter that goes home with the glossary, there is a letter that accompanies every chapter in the student text. The letter provides a model of key skills and an explanation of new vocabulary. The activities and practice exercises that follow the letter are designed to reinforce and enhance the instruction that has taken place in the classroom. In order to monitor student progress, you may wish to regard the Practice/Homework page in the *Family Involvement Activities* as a homework assignment that is returned to you for evaluation.

The use of the *Family Involvement Activities* in conjunction with the student textbook will increase family involvement in student learning and will help each student to achieve optimal success in his or her study of mathematics.

Name

Date

Dear Family,

Your child is beginning a new year in the study of mathematics. Each year, previous skills are reinforced and new skills are taught.

In an effort to provide you with important information about what your child will be learning this year, the sixth-grade Family Involvement letters describe what all sixth grade students are expected to know and be able to do. A glossary of the mathematics terms we will be learning and using this year is also included. Please keep this glossary to use as a reference throughout the school year.

As your child begins each new chapter in HARCOURT MATH, you will receive a Family Involvement Activity. Each of these activities provides instructional materials that enable older sibling, parental, or other adult tutoring of new concepts introduced in the student text; a home math activity that students can complete with family members; practice of new concepts in the form of homework; and a concept-based math game for family members to play together.

Remember that you do not have to be a mathematician to participate in your child's mathematics education. Showing your child how mathematics relates to real-world experience is an important part of his or her mathematics education—a part in which you are the major contributor.

Sincerely,

Glossary

Pronunciation Key

a	add, map	f	fit, half	n	nice, tin	p	pit, stop	yo͞o	fuse, few
ā	ace, rate	g	go, log	ng	ring, song	r	run, poor	v	vain, eve
â(r)	care, air	h	hope, hate	o	odd, hot	s	see, pass	w	win, away
ä	palm, father	i	it, give	ō	open, so	sh	sure, rush	y	yet, yearn
b	bat, rub	ī	ice, write	ô	order, jaw	t	talk, sit	z	zest, muse
ch	check, catch	j	joy, ledge	oi	oil, boy	th	thin, both	zh	vision, pleasure
d	dog, rod	k	cool, take	ou	pout, now	t͟h	this, bathe		
e	end, pet	l	look, rule	o͝o	took, full	u	up, done		
ē	equal, tree	m	move, seem	o͞o	pool, food	û(r)	burn, term		

ə the schwa, an unstressed vowel representing the sound spelled *a* in above, *e* in sicken, *i* in possible, *o* in melon, *u* in circus

Other symbols:
• separates words into syllables
ʹ indicates stress on a syllable

A

absolute value [abʹsə•lo͞ot valʹyo͞o] The distance of an integer from zero (p. 228)

acute angle [ə•kyo͞otʹ anʹgəl] an angle whose measure is greater than 0° and less than 90° (p. 320)

acute triangle [ə•kyo͞otʹ trīʹan•gəl] A triangle with all angles less than 90° (p. 332)
Example:

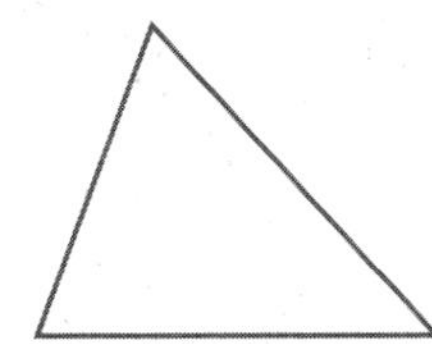

Addition Property of Equality [ə•dishʹən präʹpər•tē əv i•kwolʹə•tē] The property that states that if you add the same number to both sides of an equation, the sides remain equal (p. 290)

additive inverse [adʹə•tiv inʹvûrs] The opposite of a given number (p. 243)

adjacent angles [ə•jāʹsənt anʹgəlz] Angles that are side by side and have a common vertex and ray (p. 322)
Example:

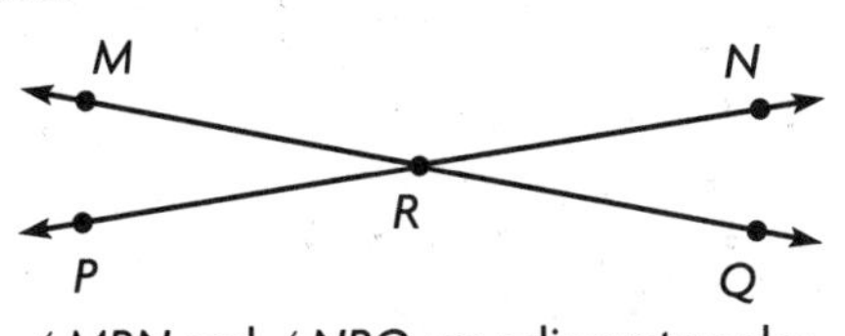

∠*MRN* and ∠*NRQ* are adjacent angles.

algebraic expression [al•jə•brāʹik ik•spreʹshən] An expression that includes at least one variable (p. 28)
Examples: $x + 5$, $3a - 4$

algebraic operating system [al•jə•brāʹik äʹpə•rā•ting sisʹtəm] A way for calculators to follow the order of operations when evaluating expressions (p. 43)

angle [anʹgəl] A figure formed by two rays with a common endpoint (p. 320)
Example:

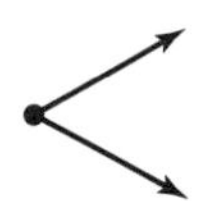

arc [ärk] A part of a circle, named by its endpoints (p. 344)
Example:

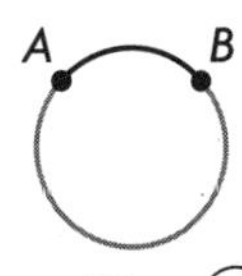

arc *AB* or $\widehat{AB}$

area [ârʹē•ə] The number of square units needed to cover a given surface (p. 494)

Associative Property [ə•sōʹshē•ā•tiv präʹpər•tē] The property that states that the way addends are grouped or factors are grouped does not change the sum or the product (p. 36)
Examples: $12 + (5 + 9) = (12 + 5) + 9$
$(9 \times 8) \times 3 = 9 \times (8 \times 3)$

axes [ak′sēz] The horizontal number line (x-axis) and the vertical number line (y-axis) on the coordinate plane (p. 570)

bar graph [bär′graf] A graph that displays countable data with horizontal or vertical bars (p. 120)

base [bās] A number used as a repeated factor (p. 40)
Example: $8^3 = 8 \times 8 \times 8$; 8 is the base.

base [bās] A side of a polygon or a face of a solid figure by which the figure is measured or named (pp. 350, 495)
Examples:

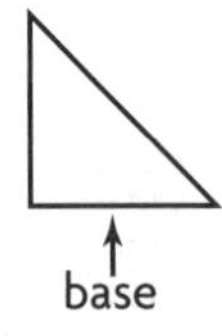

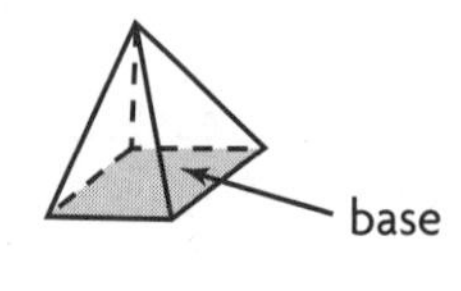

biased question [bī′əst kwes′chən] A question that leads to a specific response or excludes a certain group (p. 98)

biased sample [bī′əst sam′pəl] A sample is biased if individuals or groups from the population are not represented in the sample. (p. 98)

bisect [bī•sekt′] To divide into two congruent parts (p. 368)

box-and-whisker graph [bäks•ənd•hwis′kər graf] A graph that shows how far apart and how evenly data are distributed (p. 129)

Celsius [səl′sē•əs] A metric scale for measuring temperature (p. 301)

certain [sûr′tən] Sure to happen (p. 429)

chord [kôrd] A line segment with its endpoints on a circle (p. 344)
Example:

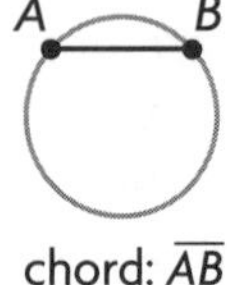

circle [sûr′kəl] A closed plane figure with all points of the figure the same distance from the center (p. 344)
Example:

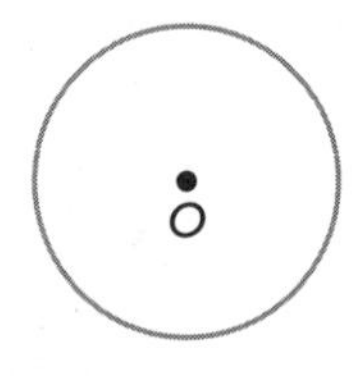

circle graph [sûr′kəl graf] A graph that lets you compare parts to the whole and to other parts (p. 122)
Example:

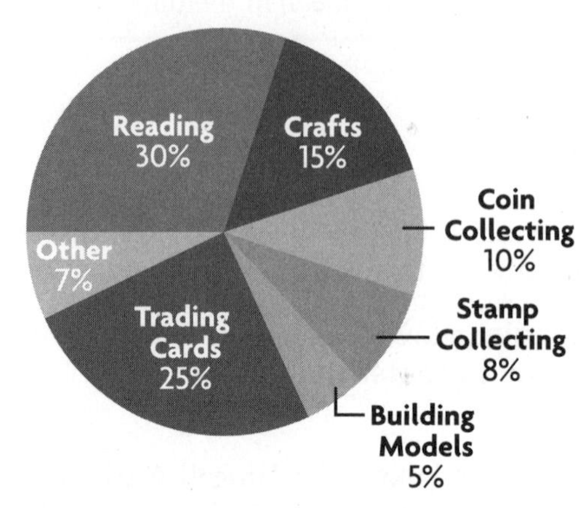

circumference [sûr•kum′fər•əns] The distance around a circle (p. 484)

clustering [klus′tər•ing] A method used to estimate a sum when all addends are about the same (p. 16)

Commutative Property [kə•myoo͞′tə•tiv prä′pər•tē] The property that states that if the order of addends or factors is changed, the sum or product stays the same (p. 36)
Examples: $6 + 5 + 7 = 5 + 6 + 7$
$8 \times 9 \times 3 = 3 \times 8 \times 9$

compensation [kom•pən•sā′shən] A mental math strategy for some addition and subtraction problems (p.37)

complementary angles [kom•plə•men′tər•ē an′gəlz] Two angles whose measures have a sum of 90° (p. 323)
Example:

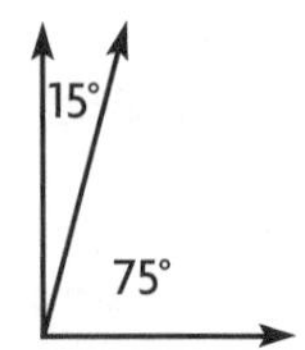

composite number [käm•pä′zət num′bər] A whole number greater than 1 that has more than two whole-number factors (p. 148)

compound event [käm′pound i•vent′] An event made of two or more simple events (p. 444)

congruent [kən•gro͞o′ənt] Having the same size and shape (p. 390)

convenience sample [kən•vēn′yənts sam′pəl] Sampling the most available subjects in the population to obtain quick results (p. 95)

coordinate plane [kō•ôr′də•nit plān] A plane formed by two perpendicular number lines called axes; every point on the plane can be named by an ordered pair of numbers. (p. 570)

corresponding angles [kôr•ə•spän′ding an′gəlz] Angles that are in the same position in different figures (p. 391)
Example:

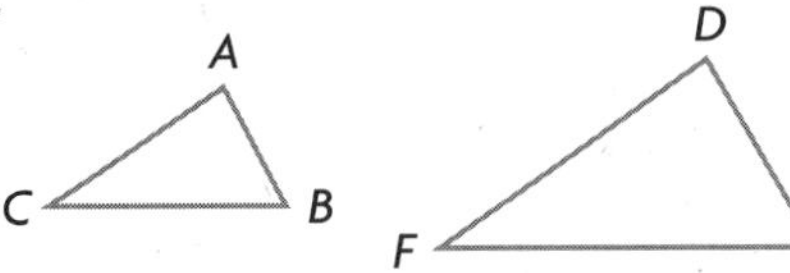

$\angle A$ and $\angle D$ are corresponding angles.

corresponding sides [kôr•ə•spän′ding sīdz] Sides that are in the same position in different plane figures (p. 391)
Example:

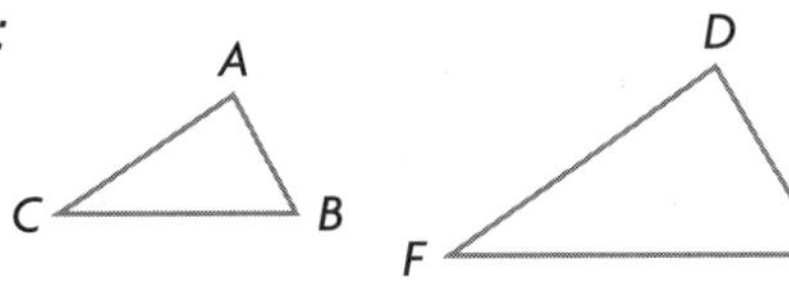

$\overline{CA}$ and $\overline{FD}$ are corresponding sides.

cube [kyo͞ob] A rectangular solid with six congruent faces (p. 350)
Example:

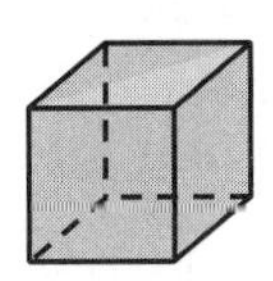

cumulative frequency [kyo͞o′myə•lə•tiv frē′kwən•sē] A running total of the number of subjects surveyed (p. 103)

D

decimal [de′sə•məl] A number with one or more digits to the right of the decimal point (p. 52)

denominator [di•nä′mə•nā•tər] The part of a fraction that tells how many equal parts are in the whole (p. 159)
Example: $\frac{3}{4}$ ←denominator

dependent events [di•pen′dənt i•vənts′] Events for which the outcome of the second event depends on the outcome of the first event (p. 448)

diameter [dī•am′ə•tər] A line segment that passes through the center of a circle and has its endpoints on the circle (p. 344)
Example:

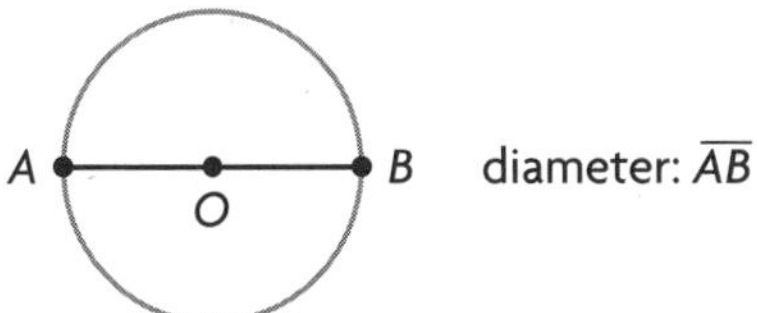

dimension [di•men′shən] The length, width, or height of a figure (p. 511)

discount [dis′kount] An amount that is subtracted from the regular price of an item (p. 418)

Distributive Property of Multiplication [di•strib′yə•tiv prä′pər•tē əv mul•tə•plə•kā′shən] The property that states that multiplying a sum by a number is the same as multiplying each addend by the number and then adding the products (p. 36)
Example: $14 \times 21 = 14 \times (20 + 1) = (14 \times 20) + (14 \times 1)$

dividend [di′və•dend] The number that is to be divided in a division problem
Example: In $56 \div 8$, 56 is the dividend.

Division Property of Equality [di•vi′zhən prä′pər•tē əv i•kwol′ə•tē] The property that states that if you divide both sides of an equation by the same nonzero number, the sides remain equal (p. 297)

divisor [di•vī′zər] The number that divides the dividend
Example: In $45 \div 9$, 9 is the divisor.

E

equally likely [ē′kwə•lē lī′klē] Having the same chance of occurring (p. 428)

equation [i•kwā′zhən] A statement that shows that two quantities are equal (p. 30)

equilateral triangle [ē•kwə•la′tə•rəl trī′an•gəl] A triangle with three congruent sides (p. 332)
Example:

P
3 cm 3 cm
R Q
3 cm

equivalent fractions [ē•kwiv′ə•lənt frak′shənz] Fractions that name the same amount or part (p. 160)

equivalent ratios [ē•kwiv′ə•lənt rā′shē•ōz] Ratios that name the same comparisons (p. 384)

estimate [es′tə•mit] An answer that is close to the exact answer and that is found by rounding, by clustering, or by using compatible numbers (p. 16)

evaluate [i•val′yo͞o•āt] Find the value of a numerical or algebraic expression (p. 28)

event [i•vent′] A set of outcomes (p. 427)

experimental probability [ik•sper•ə•men′təl prä•bə•bil′ə•tē] The ratio of the number of times an event occurs to the total number of trials or times the activity is performed (p. 436)

exponent [ik•spō′nənt] A number that tells how many times a base is used as a factor (p. 40)
Example: $2^3 = 2 \times 2 \times 2 = 8$;
3 is the exponent.

F

face [fās] One of the polygons of a solid figure (p. 493)

Example:

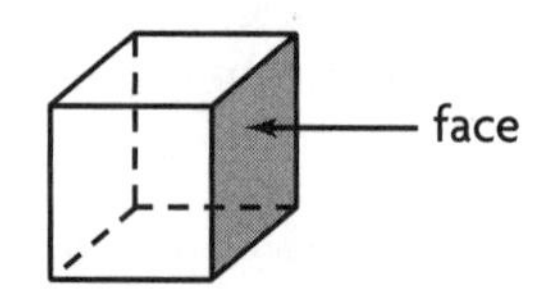

factor [fak′tər] A number that is multiplied by another number to find a product

Fahrenheit [fâr′ən•hīt] A customary scale for measuring temperature (p. 301)

formula [fôr′myə•lə] A rule that is expressed with symbols (p. 300)
Example: $A = lw$

fractal [frak′təl] A figure with repeating patterns containing shapes that are like the whole but of different sizes throughout (p. 543)

frequency table [frē′kwən•sē tā′bəl] A table representing totals for individual categories or groups (p. 103)

function [funk′shən] A relationship between two quantities in which one quantity depends on the other (p. 539)

Fundamental Counting Principle [fun•də•men′təl koun′ting prin′sə•pəl] If one event has *m* possible outcomes and a second independent event has *n* possible outcomes, then there are $m \times n$ total possible outcomes. (p. 444)

G

greatest common factor (GCF) [grā′təst kä′mən fak′tər] The greatest factor that two or more numbers have in common (p. 151)

H

height [hīt] A measure of a polygon or solid figure, taken as the length of a perpendicular from the base of the figure (p. 495)
Example:

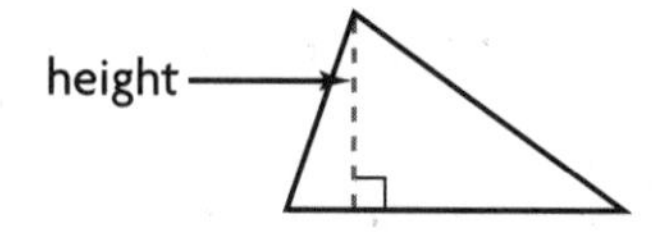

hexagon [heks′ə•gon] A six-sided polygon

histogram [his′tə•gram] A bar graph that shows the number of times data occur in certain ranges or intervals (p. 127)

hypotenuse [hī•pot′ə•n(y)o͞os′] In a right triangle, the side opposite the right angle (p. 488

Example:

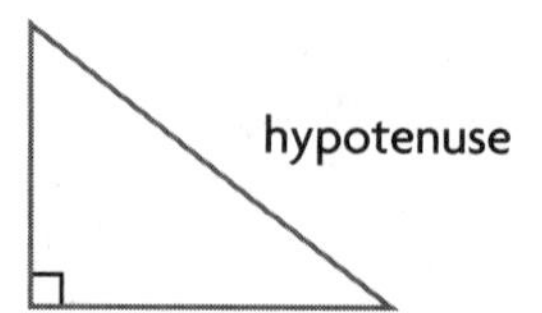

I

Identity Property of Zero [ī•den′tə•tē prä′pər•tē əv zir′ō] The property that states that the sum of zero and any number is that number (p. H2)
Example: $25 + 0 = 25$

Identity Property of One [ī•den′tə•tē prä′pər•tē əv wun] The property that states that the product of any number and 1 is that number (p. H2)
Example: $12 \times 1 = 12$

impossible [im•pos′ə•bəl] Never able to happen (p. 429)

independent events [in•di•pen′dənt i•vents′] Events for which the outcome of the second event does not depend on the outcome of the first event (p. 447)

indirect measurement [in•di•rekt′ mezh′ər•mənt] The technique of using similar figures and proportions to find a measure (p. 394)

inequality [in•i•kwäl′ə•tē] An algebraic or numerical sentence that contains the symbol $<$, $>$, $\leq$, $\geq$, or $\neq$ (p. 568)
Example: $x + 3 > 5$

integers [in′ti•jərz] The set of whole numbers and their opposites (p. 228)

isosceles triangle [ī•sä′sə•lēz trī′an•gəl] A triangle with exactly two congruent sides (p. 331)
Example:

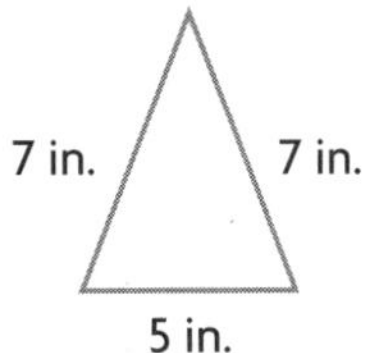

lateral faces [lat′ər•əl fās′əz] The faces in a prism or pyramid that are not bases (p. 350)

least common denominator (LCD) [lēst kä′mən di•nä′mə•nā•tər] The least common multiple of two or more denominators (p. 182)

least common multiple (LCM) [lēst kä′mən mul′tə•pəl] The smallest number, other than zero, that is a common multiple of two or more numbers (p. 150)

leg [leg] In a right triangle, either of the two sides that form the right angle (p. 488)
Example:

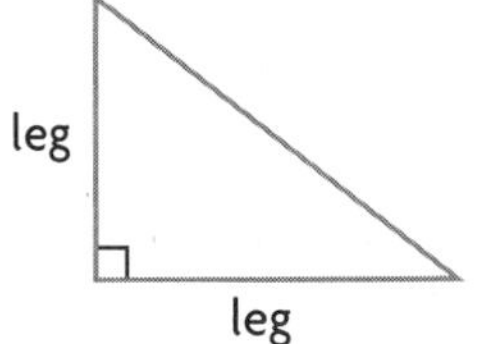

like terms [līk turmz] Expressions that have the same variable with the same exponent (p. 273)

line [līn] A straight path that extends without end in opposite directions (p. 318)
Example: ⟷

line graph [līn graf] A graph that uses a line to show how data change over time (p. 121)

line of symmetry [līn əv si′mə•trē] A line that divides a figure into two congruent parts (p. 560)

line plot [līn plät] A graph that shows frequency of data along a number line (p. 121)
Example:

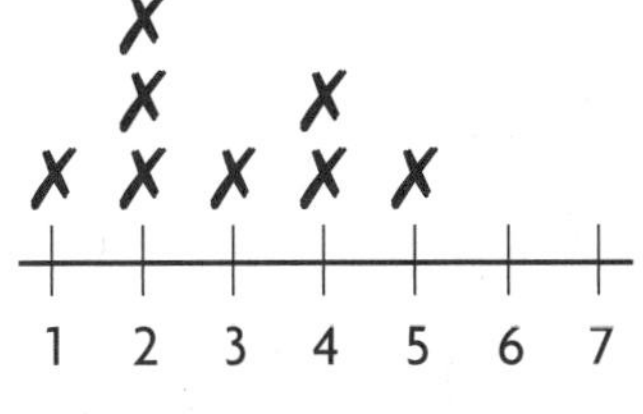

line segment [līn seg′mənt] A part of a line with two endpoints (p. 318)
Example: •———•

line symmetry [līn si′mə•trē] A figure has line symmetry if a line can separate the figure into two congruent parts. (p. 560)

lower extreme [lō′ər ik•strēm′] The least number in a set of data (p. 129)

lower quartile [lō′ər kwôr′tīl] The median of the lower half of a set of data (p. 129)

mean [mēn] The sum of a group of numbers divided by the number of addends (p. 106)

median [mē′dē•ən] The middle value in a group of numbers arranged in order (p. 106)

midpoint [mid′point] The point that divides a line segment into two congruent line segments (p. 368)

mixed number [mikst num′bər] A number represented by a whole number and a fraction (p. 164)

mode [mōd] The number or item that occurs most often in a set of data (p. 106)

multiple-bar graph [mul′tə•pəl bär′graf] A bar graph that represents two or more sets of data (p. 120)

multiple-line graph [mul′tə•pəl līn′graf] A line graph that represents two or more sets of data (p. 121)

multiple [mul′tə•pəl] The product of a given whole number and another whole number (p. 150)

Multiplication Property of Equality [mul•tə•plə•kā′shən prä′pər•tē əv i•kwol′ə•tē] The property that states that if you multiply both sides of an equation by the same number, the sides remain equal (p. 298)

negative integers [ne′gə•tiv in′ti•jərz] Integers to the left of zero on the number line (p. 228)

net [net] An arrangement of two-dimensional figures that folds to form a polyhedron (p. 356)
Example:

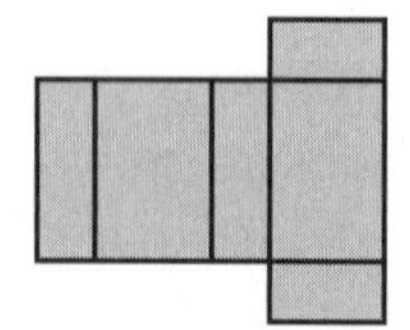

numerator [nōō′mə•rā•tər] The part of a fraction that tells how many parts are being used (p. 159)
Example: $\frac{3}{4}$ ←numerator

numerical expression [nōō•mâr′i•kəl ik•spre′shən] A mathematical phrase that uses only numbers and operation symbols (p. 28)

obtuse angle [äb•tōōs′ an′gəl] An angle whose measure is greater than 90° and less than 180° (p. 320)
Example:

obtuse triangle [äb•tōōs′ trī′an•gəl] A triangle with one angle greater than 90° (p. 332)
Example:

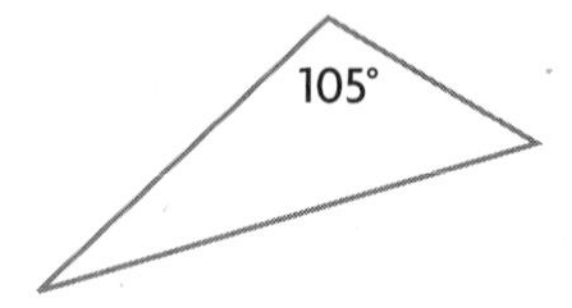

opposites [ä′pə•zəts] Two numbers that are an equal distance from zero on the number line (p. 228)

order of operations [ôr′dər əv ä•pə•rā′shənz] The process for evaluating expressions: first perform the operations in parentheses, clear the exponents, perform all multiplication and division, and then perform all addition and subtraction (p. 42)

ordered pair [ôr′dərd pâr] A pair of numbers that can be used to locate a point on the coordinate plane (p. 570)
Examples: (0,2), (3,4), ($^-$4,5)

origin [ôr′ə•jən] The point where the *x*-axis and the *y*-axis in the coordinate plane intersect, (0,0) (p. 570)

outcome [out′kəm] A possible result of a probability experiment (p. 428)

outlier [out′lī•ər] A data value that stands out from others in a set; outliers can significantly affect measures of central tendency. (p. 110)

overestimate [ō•vər•es′tə•mət] An estimate that is greater than the exact answer (p. 17)

parallel lines [pâr′ə•lel līnz] Lines in a plane that are always the same distance apart (p. 326)
Example:

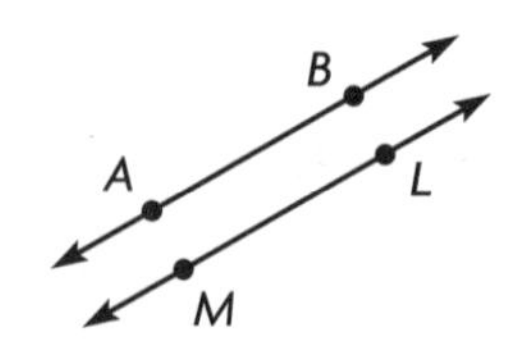

percent (%) [pər•sent′] The ratio of a number to 100; *percent* means "per hundred." (p. 60)

perimeter [pə•ri′mə•tər] The distance around a figure (p. 477)

perpendicular lines [pər•pen•dik′yə•lər līnz] Two lines that intersect to form right, or 90°, angles (p. 326)
Example:

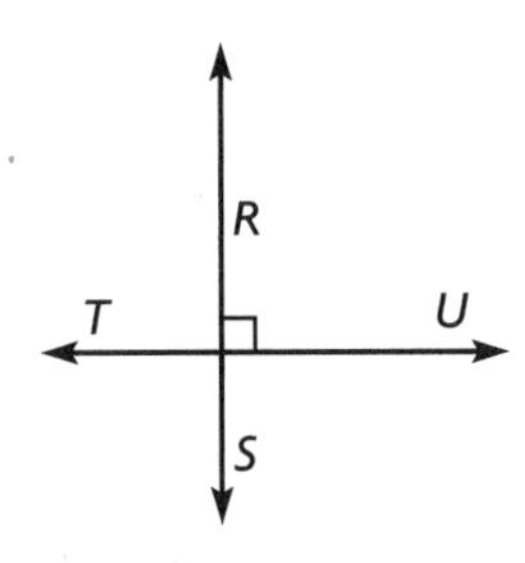

pi (π) [pī] The ratio of the circumference of a circle to its diameter; $\pi \approx 3.14$ or $\frac{22}{7}$ (p. 484)

plane [plān] A flat surface that extends without end in all directions (p. 318)

point [point] An exact location in space, usually represented by a dot (p. 318)

point of rotation [point əv rō•tā′shən] The central point around which a figure is rotated (p. 561)

polygon [pä′lē•gän] A closed plane figure formed by three or more line segments (p. 331)

polyhedron [pä•lē•hē′drən] A solid figure with flat faces that are polygons (p. 350)
Example:

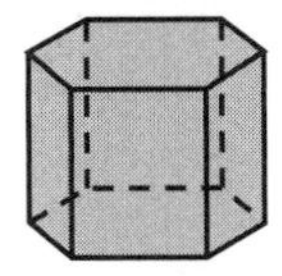

Hexagonal Prism

population [pä•pyə•lā′shən] The entire group of objects or individuals considered for a survey (p. 94)

positive integers [pä′zə•tiv in′ti•jərz] Integers to the right of zero on the number line (p. 228)

prime factorization [prīm fak•tə•ri•zā′shən] A number written as the product of all of its prime factors (p. 148)
Example: $24 = 2^3 \times 3$

prime number [prīm num′bər] A whole number greater than 1 whose only factors are 1 and itself (p. 148)

principal [prin′sə•pəl] The amount of money borrowed or saved (p. 422)

prism [priz′əm] A solid figure that has two congruent, polygon-shaped bases, and other faces that are all rectangles (p. 350)
Example:

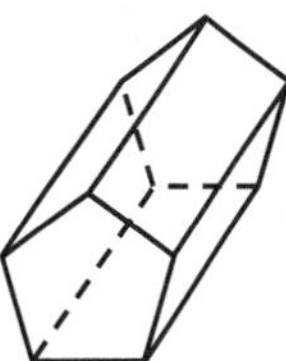

probability [prä•bə•bil′ə•tē] See *theoretical probability* and *experimental probability*

product [prä′dəkt] The answer in a multiplication problem (p. 15)

Property of Zero [prä′pər•tē əv zē′rō] The property that states that the product of any number and zero is zero (p. 35)

proportion [prə•pôr′shən] An equation that shows that two ratios are equal (p. 387)
Example: $\frac{1}{3} = \frac{3}{9}$

pyramid [pir′ə•mid] A solid figure with a polygon base and triangular sides that all meet at a common vertex (p. 351)
Example:

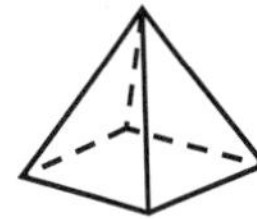

Pythagorean Theorem [pə•thag•ə•rē′ən thē′ə•rem] In any right triangle, if a and b are the lengths of the legs and c is the length of the hypotenuse, then $a^2 + b^2 = c^2$ (p. 488)

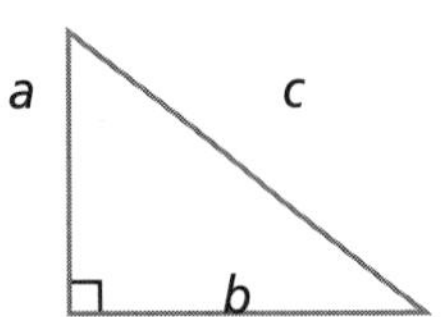

quadrants [kwäd′rənts] The four regions of the coordinate plane (p. 570)

quadrilateral [kwä•drə•lat′ə•rəl] A polygon with four sides and four angles (p. 331)

quotient [kwō′shənt] The number, not including the remainder, that results from dividing (p. 23)

R

radius [rā′dē•əs] A line segment with one endpoint at the center of a circle and the other endpoint on the circle (p. 344)
Example:

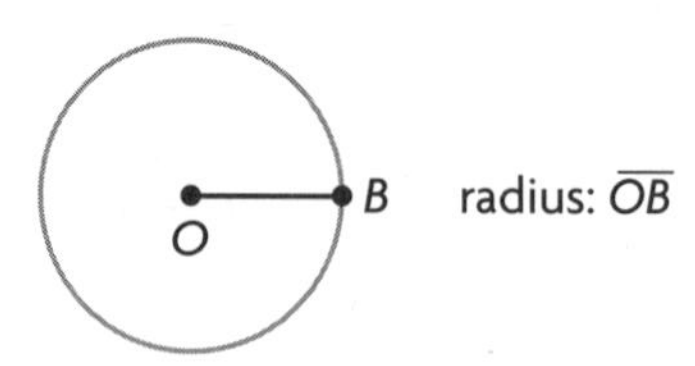

random sample [ran′dəm sam′pəl] A sample in which each subject in the overall population has an equal chance of being selected (p. 95)

range [rānj] The difference between the greatest and least numbers in a group (p. 103)

rate [rāt] A ratio that compares two quantities having different units of measure (p. 385)

ratio [rā′shē•ō] A comparison of two numbers, a and b, written as a fraction $\frac{a}{b}$ (p. 230)

rational number [ra′shə•nəl num′bər] Any number that can be written as a ratio $\frac{a}{b}$, where a and b are integers and $b \neq 0$ (p. 230)

ray [rā] A part of a line with a single endpoint (p. 318)
Example:

reciprocal [ri•sip′rə•kəl] Two numbers are reciprocals of each other if their product equals 1. (p. 209)

reflection [ri•flek′shən] A movement of a figure by flipping it over a line (p. 550)

regular polygon [reg′yə•lər pä′lē•gän] A polygon in which all sides are congruent and all angles are congruent (p. 336)
Example:

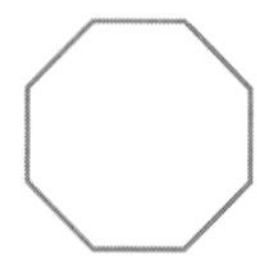

repeating decimal [ri•pēt′ing de′sə•məl] A decimal that doesn't end, because it shows a repeating pattern of digits after the decimal point (p. 169)

right angle [rīt an′gəl] An angle that has a measure of 90° (p. 320)
Example:

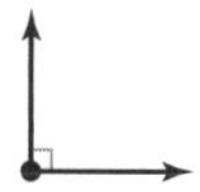

right triangle [rīt trī′an•gəl] A triangle with one right angle (p. 332)
Example:

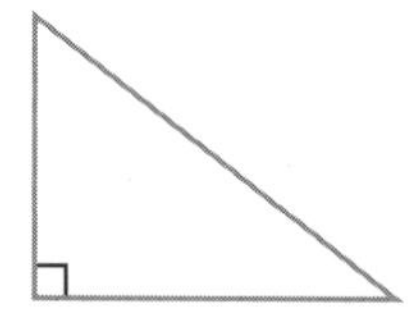

rotation [rō•tā′shən] A movement of a figure by turning it around a fixed point (p. 550)

rotational symmetry [rō•tā′shən•əl si′mə•trē] The property of a figure that can be rotated less than 360° around a central point and still be congruent to the original figure (p. 561)

S

sales tax [sālz taks] A percent of the cost of an item, added onto the item's cost (p. 420)

sample [sam′pəl] A part of a population (p. 94)

sample space [sam′pəl spās] The set of all possible outcomes (p. 428)

scale [skāl] A ratio between two sets of measurements (p. 398)

scale drawing [skāl drô′ing] A drawing that shows a real object smaller than (a reduction) or larger than (an enlargement) the real object (p. 397)

scalene [skā′lēn] A triangle with no congruent sides (p. 331)

scatterplot [skat′ər•plät] A graph with points plotted to show a relationship between two variables (p. 139)

sector [sek′tər] A region enclosed by two radii and the arc joining their endpoints (p. 344)
Example:

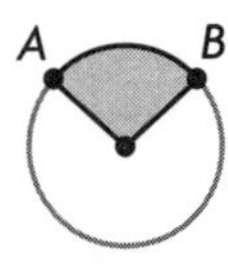

sequence [sē′kwəns] An ordered set of numbers (p. 536)

similar figures [si′mə•lər fig′yərz] Figures with the same shape but not necessarily the same size (p. 372)

simple interest [sim′pəl in′trəst] A fixed percent of the principal, paid yearly (p. 422)

simplest form [sim′pləst fôrm] The form in which the numerator and denominator of a fraction have no common factors other than 1 (p. 161)

solution [sə•lo͞o′shən] A value that, when substituted for a variable in an equation, makes the equation true (p. 30)

square [skwâr] The product of a number and itself; a number with the exponent 2 (p. 276)

square [skwâr] A rectangle with four congruent sides (p. 511)

square root [skwâr ro͞ot] One of two equal factors of a number (p. 277)

stem-and-leaf plot [stem ənd lēf plät] A type of graph that shows groups of data arranged by place value (p. 126)

straight angle [strāt an′gəl] An angle whose measure is 180° (p. 320)
Example:

Subtraction Property of Equality [sub•trak′shən prä′pər•tē əv i•kwol′ə•tē] The property that states that if you subtract the same number from both sides of an equation, the sides remain equal (p. 287)

sum [sum] The answer to an addition problem (p. 15)

supplementary angles [sup•lə•men′tə•rē an′gəlz] Two angles whose measures have a sum of 180° (p. 323)
Example:

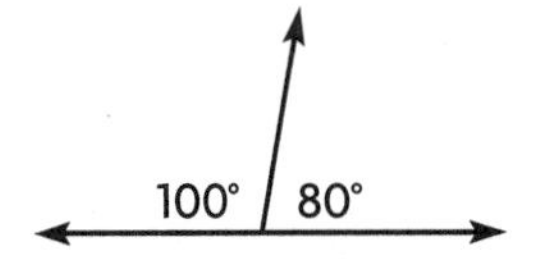

surface area [sûr′fəs âr′ē•ə] The sum of the areas of the faces of a solid figure (p. 504)

survey [sûr′vā] A method of gathering information about a group (p. 94)

systematic sample [sis•tə•ma′tik sam′pəl] A sampling method in which one subject is selected at random and subsequent subjects are selected according to a pattern (p. 95)

term [tûrm] Each number in a sequence (p. 536)

terms [tûrmz] The parts of an expression that are separated by an addition or subtraction sign (p. 273)

terminating decimal [tûr′mə•nāt•ing de′sə•məl] A decimal that ends, having a finite number of digits after the decimal point (p. 169)

tessellation [tes•ə•lā′shən] A repeating arrangement of shapes that completely covers a plane, with no gaps and no overlaps (p. 553)

theoretical probability [thē•ə•re′ti•kəl prä•bə•bil′ə•tē] A comparison of the number of favorable outcomes to the number of possible equally likely outcomes (p. 428)

transformation [trans•fər•mā′shən] A rigid transformation is a movement that does not change the size or shape of a figure (p. 550)

translation [trans•lā′shən] A movement of a figure along a straight line (p. 550)

tree diagram [trē dī′ə•gram] A diagram that shows all possible outcomes of an event (p. 444)

triangular number [trī•an′gyə•lər num′bər] A number that can be represented by a triangular array (p. 536)

unbiased sample [un•bī′əst sam′pəl] A sample is unbiased if every individual in the population has an equal chance of being selected. (p. 98)

underestimate [un•dər•es′tə•mət] An estimate that is less than the exact answer (p. 17)

unit rate [yo͞o′nət rāt] A rate that has 1 unit as its second term (p. 385)
Example: $1.45 per pound

unlike fractions [un′līk frak′shənz] Fractions with different denominators (p. 180)

upper extreme [up′ər ik•strēm′] The greatest number in a set of data (p. 129)

upper quartile [up′ər kwôr′tīl] The median of the upper half of a set of data (p. 129)

variable [vâr′ē•ə•bəl] A letter or symbol that stands for one or more numbers (p. 28)

Venn diagram [ven dī′ə•gram] A diagram that shows relationships among sets of things (p. 230)

vertex [vûr′teks] The point where two or more rays meet; the point of intersection of two sides of a polygon; the point of intersection of three or more edges of a solid figure; the top point of a cone (pp. 320, 351)
Examples:

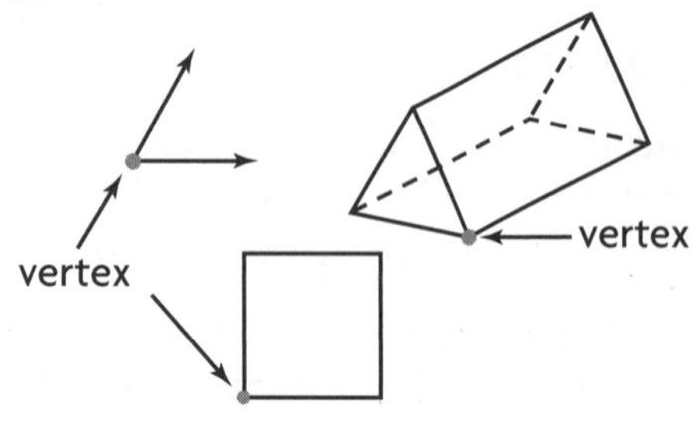

vertical angles [vûr′ti•kəl an•gəlz] A pair of opposite congruent angles formed where two lines intersect (p. 322)
Example:

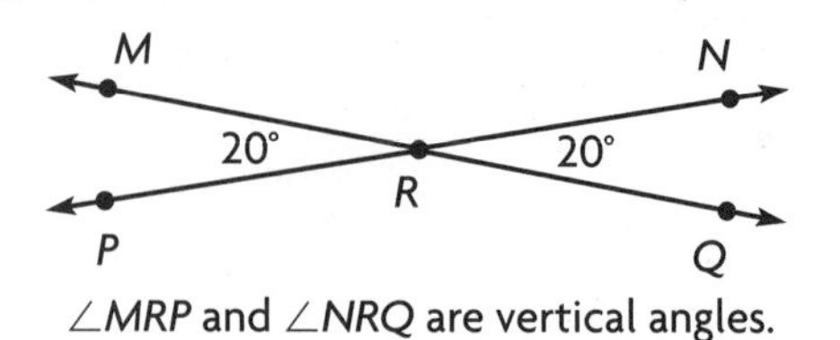

∠*MRP* and ∠*NRQ* are vertical angles.

volume [väl′yəm] The number of cubic units needed to occupy a given space (p. 512)

whole number [hōl num′bər] One of the numbers 0, 1, 2, 3, 4, ... The set of whole numbers goes on without end.

***x*-axis** [eks•ak′səs] The horizontal number line on a coordinate plane (p. 570)

***x*-coordinate** [eks•kō•ôr′də•nət] The first number in an ordered pair; it tells the distance to move right or left from (0,0).

***y*-axis** [wī•ak′səs] The vertical number line on a coordinate plane (p. 570)

***y*-coordinate** [wī•kō•ôr′də•nət] The second number in an ordered pair; it tells the distance to move up or down from (0,0).

Name

GRADE 6

Date

Chapter 1

WHAT WE ARE LEARNING

Whole Number Applications

VOCABULARY

Here are some of the vocabulary words we use in class:

Numerical expression A mathematical phrase that uses only numbers and operation symbols

Variable A letter or symbol that stands for one or more numbers

Algebraic expression An expression that includes at least one variable

Evaluate The process of finding the value of a numerical expression or an algebraic expression

Equation A statement showing that two quantities are equal

Dear Family,

Your child will be working on estimating, numerical and algebraic expressions, and solving equations. Discuss with your child how you use estimating and mental math on a day-to-day basis to solve problems.

Problem: You need gasoline, but you only have $10.00. Gasoline costs $1.25 per gallon. How many gallons can you buy?

Ask questions such as these as you work together.

- **What are you asked to find?** Your child might say: How many gallons of gas can I buy with $10.00?
- **What do you need to know to find the answer?** Your child might say: I need to know how much money I have and how much gasoline costs.
- **What kind of mathematical operation can you use to solve it?** I can multiply the number of gallons I buy by $1.25 to get $10.00, the amount I will spend.
- **How can you write that idea as an algebraic equation?** Your child might answer: $\$1.25 \times n = \$10.00$ where n is the number of gallons that I can purchase.
- **How do you solve the equation using mental math?** Your child might say: I try values for the variable n until I find a value that makes the equation true. The solution is 8 because $\$1.25 \times 8 = \$10.00$.

Use this model and the exercises that follow this page to help your child practice estimating and computing with whole numbers, evaluating expressions, and using mental math to compute and to solve equations.

Sincerely,

They Are The Same!

HOME ACTIVITY

Play this game to find equal expressions.

Directions:

1. Cut out the cards at the bottom of the page.
2. Mix up the cards.
3. Play a game of concentration. A pair is two cards that name the same number when the expressions are evaluated.
4. The first player turns over two cards. If the two cards make a pair, the player keeps them. If they do not make a pair, the player puts the cards back in the same place.
5. Players take turns turning over two cards and trying to make pairs.
6. The game is over when all of the cards are gone.
7. The person with the most pairs at the end of the game is the winner.

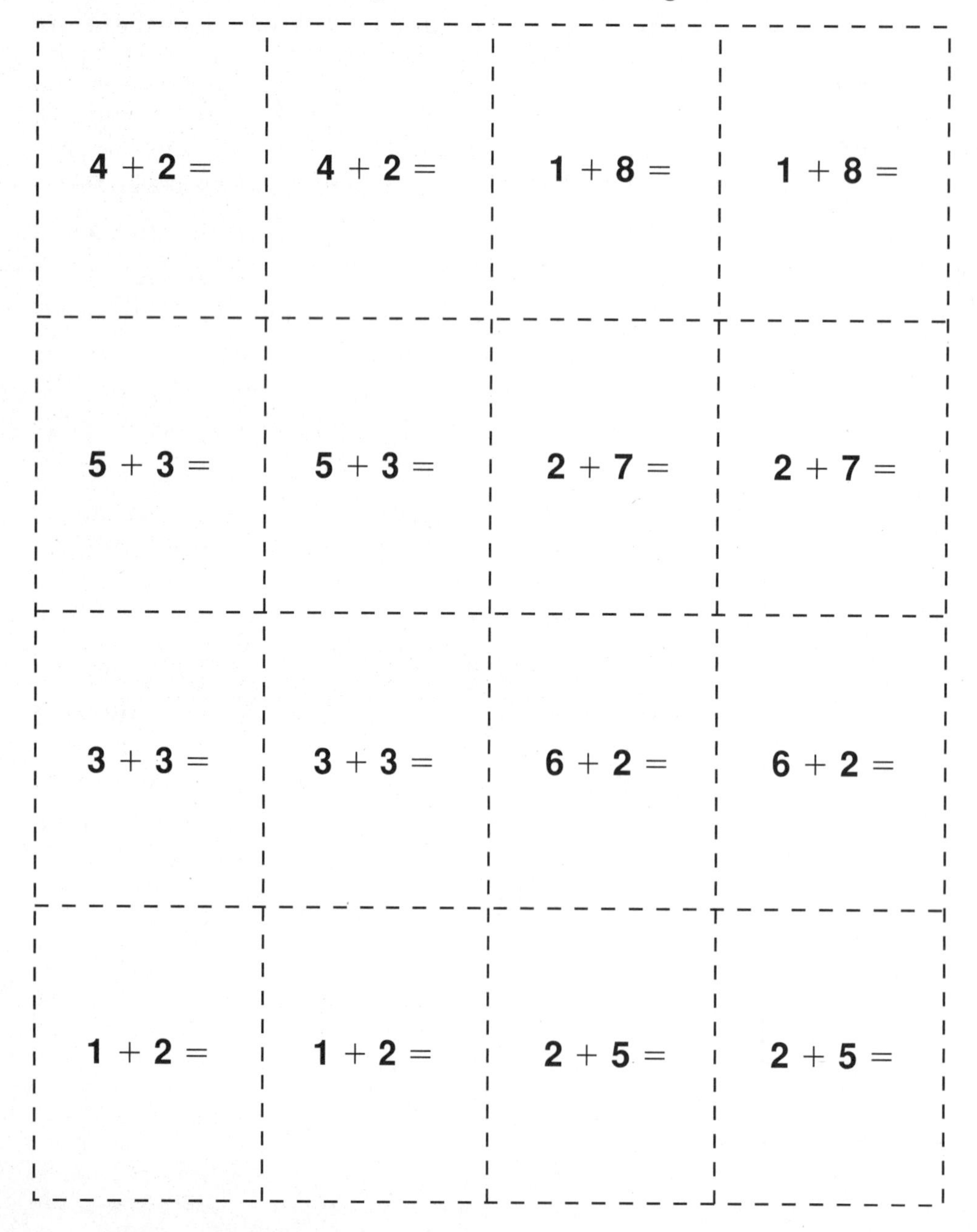

$4 + 2 =$	$4 + 2 =$	$1 + 8 =$	$1 + 8 =$
$5 + 3 =$	$5 + 3 =$	$2 + 7 =$	$2 + 7 =$
$3 + 3 =$	$3 + 3 =$	$6 + 2 =$	$6 + 2 =$
$1 + 2 =$	$1 + 2 =$	$2 + 5 =$	$2 + 5 =$

Name ______________________________

Whole Number Applications

Estimate the sum, difference, product, or quotient.

1. 23,923 + 97,148 + 56,488 __________
2. 6,827 + 7,623 + 7,444 __________
3. 27,777 – 14,156 __________
4. 9,921 – 7,099 __________
5. 187 × 23 __________
6. 521 × 37 __________
7. 897 ÷ 34 __________
8. 5,047 ÷ 48 __________

Write an algebraic expression for the word expression.

9. Three times a weight, w _____
10. A pie, p, divided into 4 equal pieces _____
11. 6 feet less than the height of a tree, t _____
12. Three hours more than the total hours worked on Monday, m _____

Evaluate each expression.

13. $a - 75$, for $a = 220$ _____
14. $d \times 40$, for $d = 10$ _____
15. $240 + b$, for $b = 80$ _____
16. $c \div 20$, for $c = 300$ _____

Solve each equation.

17. $a + 12 = 15$ _____
18. $b \div 12 = 4$ _____
19. $c \times 23 = 460$ _____
20. $a - 7 = 15 + 5$ _____

Answers: For 1-8, possible estimates are given. **1.** 177,000; **2.** 22,000; **3.** 14,000; **4.** 3,000; **5.** 4,000; **6.** 20,000; **7.** 30; **8.** 100; **9.** $3w$; **10.** $p \div 4$; **11.** $t - 6$; **12.** $m + 3$; **13.** 145; **14.** 400; **15.** 320; **16.** 15; **17.** $a = 3$; **18.** $b = 48$; **19.** $c = 20$; **20.** $a = 27$

Family Fun

Let's Bake It

MATH GAME

Chocolate Chip Cookies

$\frac{2}{3}$ cup shortening

$\frac{1}{2}$ cup granulated sugar

$\frac{1}{2}$ cup brown sugar

1 egg

1 teaspoon vanilla

$1\frac{1}{2}$ cups flour

$\frac{1}{2}$ teaspoon baking soda

$\frac{1}{2}$ teaspoon salt

$\frac{1}{2}$ cup chopped nuts

1 package semi-sweet chocolate pieces

Heat oven to 375°. Blend dry ingredients. Mix wet ingredients thoroughly. Stir in dry ingredients. Mix in nuts and chocolate. Drop rounded teaspoonfuls 2 inches apart onto ungreased baking sheet. Bake 8 to 10 minutes. Cool slightly. Makes 5 dozen cookies.

Tell your child that you want to use the recipe to make as many cookies as possible. You have lots of all of the ingredients except vanilla. Only 1 tablespoon (3 teaspoons) of vanilla is left in the bottle.

1. Ask your child to write an algebraic equation that could be used to find
   a. the number of dozens of cookies that can be made with the available vanilla.
   b. the number of cookies in the number of dozens you found in part a.
2. Ask your child to solve the equations in problem 1 and explain the procedures used.

Possible equations are given. **1a.** $3 \times 5 = d$, where d is the number of dozens of cookies; **1b.** $3 \times 5 \times 12 = n$, where n is the number of cookies; **2.** 15 dozen cookies; 180 cookies; explanations will vary

HARCOURT MATH

Name

GRADE 6

Date

Chapter 2

WHAT WE ARE LEARNING

Operation Sense

VOCABULARY

Here are some of the vocabulary words we use in class:

Compensation A mental math strategy you can use for some addition and subtraction problems

Exponent An exponent shows how many times a number is used as a factor

Base The base is the number that an exponent shows is used as a factor

Algebraic operating system Some calculators use an algebraic operating system so they automatically follow the order of operations.

Dear Family,

Your child is learning about using mental math strategies to find sums, differences, products, and quotients; using exponents to represent numbers; and using the order of operations.

This is how your child is learning to use properties and mental math to find sums.

Ask questions such as these as you work together.

- Use the Commutative Property.

$$\begin{aligned} 54 + 52 + 6 &= 54 + 6 + 52 \\ &= 60 + 52 \\ &= 112 \end{aligned}$$

Can you explain the Commutative Property and the Associative Property to me? Your child might say: With the Commutative Property I can change the order of the addends and get the same sum.

- Use the Associative Property.

$$\begin{aligned} (54 + 52) + 6 &= 54 + (52 + 6) \\ &= 54 + 58 \\ &= 112 \end{aligned}$$

The Associative Property lets me group addends in any way and get the same sum.

- Use compensation.

$$\begin{aligned} 54 + 52 &= (54 + 6) + (52 - 6) \\ &= 60 + 46 \\ &= 106 \end{aligned}$$

Can you explain the strategy of compensation? Your child might respond: When I'm adding, I can change one addend to a multiple of ten and then adjust the other addend by subtracting the same number to keep the balance. When I use compensation to subtract, I have to do the same thing to each number.

Why might you use these properties? Your child might reply: When I use mental math, the properties help me to find the answer more easily.

Your child has learned this about exponents.

Step 1: To find the value of a number expressed by a base and an exponent, use the base as a factor the number of times indicated by the exponent.

$$3^4 = 3 \times 3 \times 3 \times 3$$
$$= 81$$

Step 2: To represent a number using the base and an exponent, find the equal factors. Write the factor as the base and the number of times it is used as the exponent.

$$81 = 3 \times 3 \times 3 \times 3$$
$$= 3^4$$

What is a factor and what is a base? Your child might answer: Factors are numbers that are multiplied to get a product. An exponent shows how many times a number called the base is used as a factor.

Can you show me how to find the value of 7^5? Your child might answer: I use 7 as a factor 5 times. $7 \times 7 \times 7 \times 7 \times 7 = 16{,}807$.

How would you write 512 using an exponent and the base 8? Your child might answer: $8 \times 8 \times 8 = 512$, so the base is 8 and the exponent is 3: 8^3.

As you work with your child, talk about math to help build confidence and understanding.

Sincerely,

Name ______________________________

PRACTICE/HOMEWORK

Operation Sense

Use mental math to find the values.

1. 25×8 ________
2. $83 + 42$ ________
3. $(33 + 19) + 7$ ________
4. $23 + 42 + 61$ ________

Name the missing reason for each step.

5. $70 \times 3 = (7 \times 10) \times 3$ 70 means 7×10
   $= 7 \times (10 \times 3)$ Associative Property
   $= 7 \times (3 \times 10)$ ____________________
   $= (7 \times 3) \times 10$ ____________________
   $= 21 \times 10$ ____________________
   $= 210$ ____________________

Write the equal factors. Then find the value.

6. 6^5
7. 42^1
8. 62^2
9. 10^6
10. 1^8

Write in exponent form.

11. $17 \times 17 \times 17 \times 17$ ________
12. $3 \times 3 \times 3 \times 3 \times 3$ ________
13. $n \times n \times n$ ________

Evaluate the expressions.

14. $26 - (32 - 16) + (23 - 21)^2$ ________
15. $(5^2 - 4^2) + (6 \times 4) + 2$ ________

Evaluate the expression for $a = 3$ and $b = 8$.

16. $30 + a + 12$ ________
17. $(11 - b) + 3$ ________
18. $a^2 + 1 \times (4 + 5)$ ________

Answers: **1.** 200; **2.** 125; **3.** 59; **4.** 126; **5.** Commutative Property, Associative Property, $7 \times 3 = 21$, $21 \times 10 = 210$; **6.** $6 \times 6 \times 6 \times 6 \times 6$; 7,776; **7.** 42; 42; **8.** 62×62; 3,844; **9.** $10 \times 10 \times 10 \times 10 \times 10 \times 10$; 1,000,000; **10.** $1 \times 1 \times 1 \times 1 \times 1 \times 1 \times 1 \times 1$; 1; **11.** 17^4; **12.** 3^5; **13.** n^3; **14.** 14; **15.** 35; **16.** 45; **17.** 6; **18.** 90

Family Fun

Convert the value of the following expressions into letters to read the message. HINT: Pélé said it.

A = 12	B = 5	C = 6	D = 14	E = 8	F = 15	G = 13	H = 2
I = 9	J = 20	K = 51	L = 17	M = 18	N = 7	O = 0	P = 20
Q = 19	R = 3	S = 10	T = 1	U = 5	V= 50	W = 4	X = 16
Y = 21	Z = 62						

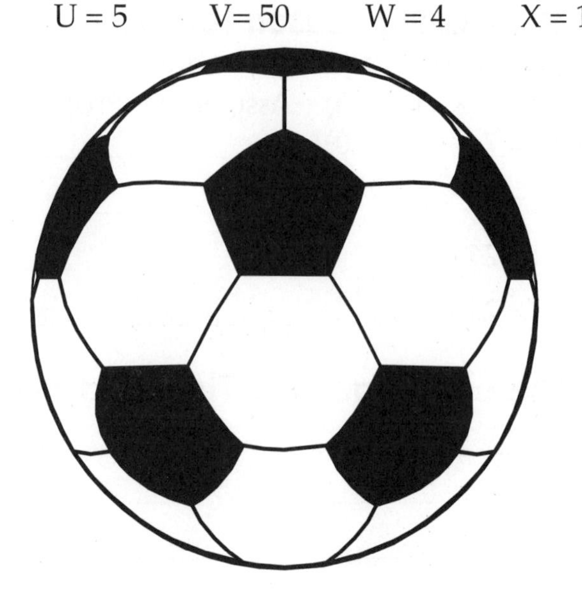

1. _____ $(6 + 12) - 9$
2. _____ 2^2
3. _____ $(5 \times 2) + 2$
4. _____ 10^1
5. _____ $25 = n^2$, $n =$ _____
6. _____ $(8 + 12) - (12 + 8)$
7. _____ $(13 - 8) \times 2 - 7$
8. _____ $(10^1 - 3) \times 1^2$
9. _____ $5^2 - 10$
10. _____ $(10 - 8) - 2$
11. _____ What is the exponent in the expression 11^3?
12. _____ $17 + 23 - 16 - 7 \times 2$
13. _____ $45 \times 2 + 7 + 3 - 10^2$
14. _____ What is the base in the expression 6^2?
15. _____ $48 \div 4 \div 2$
16. _____ One less than 3^2
17. _____ $15 - 4 \times 3$

Answer: I was born for soccer.

HARCOURT MATH

GRADE 6

Chapter 3

WHAT WE ARE LEARNING

Decimal Concepts

VOCABULARY

Here is a vocabulary word we use in class:

Percent "Per hundred" or the ratio of a number to 100. The symbol used to write a percent is %.

Name

Date

Dear Family,

Your child is studying place value with decimals; estimation of decimal sums, differences, products, and quotients; and changing decimals to percents and percents to decimals.

This is how your child is learning to use place value to express, compare, and order decimals.

O	.	T	H
6	.	3	7
6	.	3	
2	.	9	5
0	.	2	5
5	.	5	

To compare decimals, use place value.

Compare the digits from left to right.

Which is greater, 6.37 or 6.3?

Same number of ones digits.	$\underline{6}.37$	$\underline{6}.3$
Same number of tenths digits.	$6.\underline{3}7$	$6.\underline{3}$
Add a zero.		
Compare hundredths digits.	$6.3\underline{7}$	$6.3\underline{0}$
7 is greater than 0.		
So, $6.37 > 6.3$ and $6.3 < 6.37$.		

Your child is learning to use the same methods to estimate sums, differences, products, and quotients of decimals as were used for estimating with whole numbers.

3.69 + 3.79 + 4.07 — The three addends cluster around 4, so multiply 4 by 3.
$3 \times 4 = 12$
The sum is about 12.

3.69 x 3.79 — Round to the nearest one.
Then multiply. $4 \times 4 = 16$
The product is about 16.

$23.4 \div 6.3$ — Use compatible numbers. Then divide. $24 \div 6 = 4$
The quotient is about 1.

This is one way your child is learning to write percents as decimals and decimals as percents.

To write a percent as a decimal	Write 27% as a decimal.
1. Express the percent as a fraction with 100 as the denominator.	First, write 27% as a fraction: $\frac{27}{100}$
2. Write the fraction as a decimal.	Then, write $\frac{27}{100}$ as a decimal: 0.27.
To write a decimal as a percent	Write 0.71 as a percent.
1. Express the decimal as a fraction with 100 as the denominator.	First write 0.71 as a fraction: $\frac{71}{100}$
2. Write the fraction as a percent.	Then write $\frac{71}{100}$ as a percent: 71%.

As you work with your child, talk about math to help build confidence and understanding.

Sincerely,

Name ______________________

PRACTICE/HOMEWORK

Decimal Concepts

Write the value of the underlined digit.

1. $3.10\underline{9}2$ __________
2. $0.053\underline{7}$ __________
3. $8.\underline{6}82$ __________
4. $0.0\underline{2}51$ __________

Write the number in expanded form.

5. 0.00301 ______________________________
6. 9.128 ______________________________
7. 11.0643 ______________________________
8. 159.07 ______________________________

Compare the numbers. Write <, >, or = in the blank.

9. 8.09 __ 8.094
10. 331.47 __ 321.47
11. 7.26 __ 7.263

Write the numbers in order from least to greatest.

12. 36.79, 39.76, 39.67, 37.96 ______________________
13. 0.004, 0.040, 0.400, 0.044 ______________________

Estimate.

14. 7.8 + 8.2 + 8.03 _____
15. 53.9 × 5 _____
16. 258.8 – 37.9 _____
17. 55.21 × 9.8788 _____

Write the decimal and percent for the shaded part.

18. 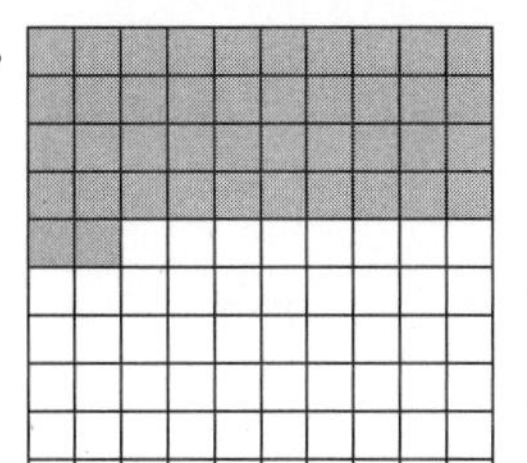 _____ _____

19. 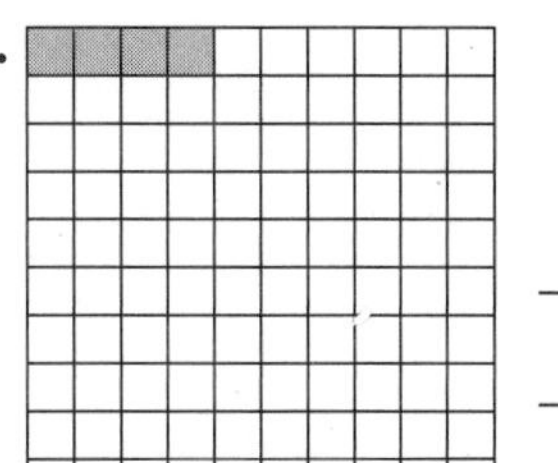 _____ _____

20. 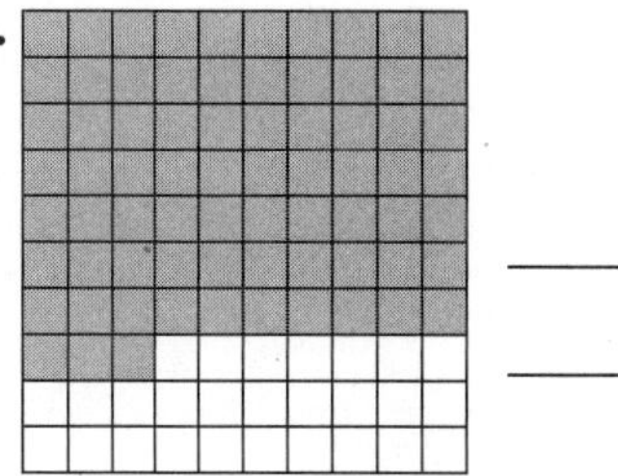 _____ _____

Write the corresponding decimal or percent.

21. 90% _____
22. 0.47 _____
23. 7% _____
24. 0.08 _____

Answers: 1. 9 thousandths; **2.** 7 ten thousandths; **3.** 6 tenths; **4.** 2 hundredths; **5.** 0.003 + 0.00001; **6.** 9 + 0.1 + 0.02 + 0.008; **7.** 10 + 1 + 0.06 +0.004 + 0.0003; **8.** 100 + 50 + 9 + 0.07; **9.** <; **10.** >; **11.** <; **12.** 36.79, 37.96, 39.67, 39.76; **13.** 0.004, 0.040, 0.044, 0.400; For **14-17**, possible estimates are given. **14.** 24; **15.** 250; **16.** 220; **17.** 550; **18.** 0.42, 42%; **19.** 0.04, 4%; **20.** 0.73, 73%; **21.** 0.9, or 0.90; **22.** 47%; **23.** 0.07; **24.** 8%

MATH GAME

Family Fun CONCENTRATION

Directions: Cut out the cards below and place them face down in 4 rows and 4 columns. With two guesses, try to match the expression with its value. When you have found a pair, remove it from the board. If you do not find an expression and its value, return the cards to their original spots.

10.2	ten and 2 tenths	thirty six and seventy-nine hundredths	36.79
$8.24 > 8.239$	$8.239 < 8.24$	Numbers listed from greatest to least	8.000, 0.809, 0.800, 0.098
about 11	5.921 + 3.1112 + 1.59876	about 230	22.634×9.798
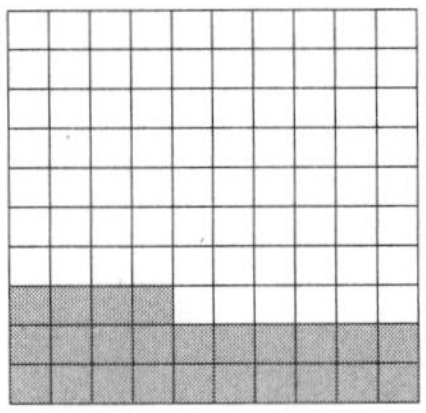	24%, 0.24	0.68	68%

Name

Date

WHAT WE ARE LEARNING

Decimal Operations

VOCABULARY

Here are some of the vocabulary words we use in class:

Decimal A number that uses place value and a decimal point to show tenths, hundredths, thousandths, and so on

Expression A mathematical phrase that combines operations, numerals, and/or variables to name a number

Dear Family,

In this chapter, your child is adding, subtracting, multiplying, and dividing with decimals and evaluating expressions and equations with decimals.

This is how your child is learning to add and subtract with decimals.

	Add: 5.43 + 7 + 0.588	Subtract: 41.4 – 7.0388
Step 1 Estimate. • Round to the nearest whole number. • Find the estimated answer.	5.43 → 5 7 → 7 + 0.588 → + 1 13	41.4 → 41 – 7.0388 → – 7 34
Step 2 Compute. • Align the decimal points. • Use zeros as place holders. • Place the decimal point. • Compute.	5.430 7.000 + 0.588 13.018	41.4000 – 7.0388 34.3612
Step 3 Compare. • Compare the answer to your estimate. • Is your answer reasonable?	13.018 is close to 13. The answer is reasonable.	34.3612 is close to 34. The answer is reasonable.

This is how your child is learning to multiply with decimals.

	Multiply: 47.85 x 3.6	
Step 1 Estimate.		
• Round to the nearest whole number.	47.85 × 3.6	48 × 4
• Find the estimated answer.		192
Step 2 Compute.		
• Multiply as with whole numbers.	47.85 × 3.6	2 places 1 place
• Place the decimal point in the product by adding decimal places in the factors or by using the estimated answer.	28710 143550 172.260	 3 places
Step 3 Compare.		
• Compare the answer to your estimate.	172.260 is close to 192.	
• Is your answer reasonable?	The answer is reasonable.	

As you work with your child, talk about math to help build confidence and understanding.

Sincerely,

Name ____________________

PRACTICE/HOMEWORK

Decimal Operations

Add or subtract.

1. 4.61 + 7.8 ______
2. 20.99 – 13.05 ______
3. 7.9673 – 6.8812 ______
4. 8.23 + 3.94 + 11.513 ______
5. 13.601 – 11.310 ______
6. 8 - 0.773 ______
7. 9 + 2.14 + 5.36 ______
8. 8.05 – 1.622 ______
9. $15.45 – $12.25 ______

Tell the number of decimal places there will be in the product.

10. 37.1 × 7.9 ______
11. 4.08 × 5.13 ______
12. 2.7 × 2.07 ______
13. 0.007 × 3.5 ______

Multiply.

14. 7 × 2.2 ______
15. 5 × 2.89 ______
16. 35.3 × 10.12 ______
17. 3.49 × 4.74 ______

Rewrite the problem so that the divisor is a whole number.

18. 17.8 ÷ 0.42 ______ 19. 895 ÷ 6.23 ______ 20. 30.07 ÷ 0.67 ______

Find the quotient.

21. 18.6 ÷ 6 _____ 22. 4.88 ÷ 0.4 _____ 23. 7.44 ÷ 1.2 _____ 24. 33.872 ÷ 7.3 _____

Evaluate each expression.

25. 7.5 × a for a = 19 ______ 26. 7.9 + 8.13 + f for f = 2.4 ______

Answers: 1. 12.41; **2.** 7.94; **3.** 1.0861; **4.** 23.683; **5.** 2.291; **6.** 7.227; **7.** 16.5; **8.** 6.428; **9.** $3.20; **10.** 2; **11.** 4; **12.** 3; **13.** 4; **14.** 15.4; **15.** 14.45; **16.** 357.236; **17.** 16.5426; **18.** 1780/42; **19.** 89500/623; **20.** 3007/67; **21.** 3.1; **22.** 12.2; **23.** 6.2; **24.** 4.64; **25.** 142.5; **26.** 18.43

MATH GAME

Family Fun

BURGER MATH

Your fast food company plans to sell 2 million hamburgers this year. As the CFO, the chief financial officer, you must figure out if this goal is profitable.

Here are most of the ingredients needed for a basic hamburger:

113.5 g Beef	0.19 g Salt
2.1 mL Ketchup	2.76 mL Special Sauce
1.5 mL Mustard	27.3 g Instant Bun Mix

Here are the questions you need to answer before the company can move forward. Since you must report your findings to the Board of Directors, you will need to write down each step you take so you can explain your decision. Write down your steps as you figure out the answers.

- If the average cow yields 175.5 kg of ground beef, how many cows will the company need to reach its goal of 2 million hamburgers?
- If a tank truck holds 10 m^3 (cubic meters), how many truck loads of ketchup will be needed?
- How many truck loads of mustard will be needed?
- How many tons of salt should the company order? (1 lb = 454 g)
- If your company sold hamburgers for $1.29 and they cost $0.89 to make, what would your annual profit be?
- What could you do to increase your profits? Explain.

Answers: 1. 1,294; 2. 0.42; 3. 0.3; 4. 0.42; 5. $800,000.00

Name

Date

WHAT WE ARE LEARNING

Collecting and Organizing Data

VOCABULARY

Here are some of the vocabulary words we use in class:

Survey A method of gathering information about a group

Population The entire group of individuals or objects that could be a part of a survey

Sample A part of the population selected for a survey

Random sample A survey in which every individual or object in the population has an equal chance of being selected

Dear Family,

Your child is learning to identify different kinds of samples and to determine if surveys are biased.

This is how to recognize bias in the sample or in the survey questions.

The local boys and girls club is conducting a survey about teens' favorite vacation locations.

Biased	Unbiased
Randomly survey 15 out of 100 girls	Randomly survey 15 out of 100 teenagers
"Do you agree with the members of the Pep squad that the beach is the best spot for a vacation?"	"Where is your favorite place to go on vacation?"

This is how your child is learning to work with and analyze measures of central tendency.

Find the mean, median, and mode for 25, 38, 43, 27, and 38.

Mean: $(25 + 38 + 43 + 27 + 38) \div 5 = 34.2$
Median: 25 27 [38] 38 43; 38
Mode: 38

Ask questions such as these as you work together.

Explain how to find the mean, median, and mode.

Your child might respond: I find the mean by adding a group of numbers and dividing by the number of addends. The median is the number in the middle. The mode is the number that occurs most often.

How do you know which of the three measures of central tendency to use?

Your child might explain: I look for the central tendency that is closest to the data. Often it is the mean, but sometimes it is the median or the mode.

Determine which measure of central tendency is most useful to describe the data above.

The mean would be the most useful central tendency.

Systematic sample A survey in which an individual is randomly selected and then others are selected by using a pattern

Convenience sample A survey in which the most available individuals in the population are selected to obtain results quickly

Frequency table A table that shows the total for each category or group

Cumulative frequency A record that shows a running total of people's responses

Outlier A data value that stands out from other data values in a set

How do you determine if a conclusion is valid?

Your child might reply: I ask four questions. Who are the people I am studying? Was the sample selected from the correct population? Is the question unbiased and fair? Was the sample randomly selected? If the answer is yes to all of these questions, then the conclusion is valid.

Students should recognize that additional data may affect the central tendencies.

Runs Scored by the Titans

Week	1	2	3	4
Runs	7	5	4	5

Mean: 5.25 Mode: 5 Median: 5

Runs Scored by the Titans

Week	1	2	3	4	5	6
Runs	7	5	4	5	4	4

Mean: 4.8 Mode: 4 Median: 4.5

As you work with your child, talk about math to help build confidence and understanding.

Sincerely,

Name ______________________

Collect and Organize Data

Identify the type of sample.

1. A company president randomly selects an employee and then surveys every tenth employee on the company payroll.

   ____

2. Each citizen of the village is assigned a number. One hundred of the citizens are randomly selected by computer.

   ____

Pollsters are surveying teens' about their favorite stores at the mall. Tell whether the sampling method is biased or unbiased.

3. Randomly survey 1 out of every 20 teenagers at the middle school and high school. ______________

4. Randomly sample 10 students who walk to the middle school.

   ____

Favorite Dog	Tally Marks	Frequency	Cumulative Frequency
Golden Retriever	|||| |||| |||| |		
German Shepherd	|||| ||||		
Labrador	|||| |||| |||		
Chihuahua	|||| ||		

Complete the table below.

Use the frequency data in the table above to answer these questions.

5. What is the mean? ______________

6. What is the median? ______________

7. What is the mode? ______________

8. What is the range? ______________

Answers: 1. systematic; 2. random; 3. unbiased; 4. biased; Table: Column 3: 16, 10, 13, 7; Column 4: 16, 26, 39, 46; 5. 11.5; 6. 11.5; 7. None; 8. 9

MATH GAME

Family Fun CRAZY CROSSWORD

ACROSS

1. The entire group of individuals that could be a part of a survey
2. Mean, median, and mode are measures of _______ tendency.
3. You find the ______ by adding a group of numbers and dividing by the number of addends.
4. The mean is the measure of central _____ that is the average of a group of numbers.
5. The type of sample in which every individual in the population has an equal chance of being selected.

DOWN

6. A data value that stands out from the others in the set.
7. A method of gathering information about a group.
8. The part of the group to be surveyed is a _____.
9. In the set of numbers 17, 21, 25, 33, 33, the _____ is 33.
10. A _____ table shows the total for each category or group.
11. One method for recording data is a line _____.
12. Information about a group.
13. The middle number in a data set.

Answer: 1. population; 2. central; 3. mean; 4. tendency; 5. random; 6. outlier; 7. survey; 8. sample; 9. mode; 10. frequency; 11. plot; 12. data; 13. median

HARCOURT MATH

GRADE 6

Chapter 6

WHAT WE ARE LEARNING

Graphing Data

VOCABULARY

Here are some of the vocabulary words we use in class:

Multiple-bar graph A bar graph that shows two or more sets of data on the same graph

Multiple-line graph A line graph that shows two or more sets of data on the same graph

Stem-and-leaf plot A way to organize data when you want to see each item in the data

Histogram A bar graph that shows the frequency, or number of times, data occur within intervals

Box-and-whisker graph A graph that shows how far apart and how evenly data are distributed

Name

Date

Dear Family,

Your child is analyzing graphs and making stem-and-leaf plots and box-and-whisker graphs.

To make a stem-and-leaf plot of the data, follow the steps below.

Population Density of Selected South American Countries per Square Mile
30, 25, 17, 46, 45, 72, 68, 9, 77, 61, 27, 44, 56

Step 1

First, group data by tens digits. Then, order data from least to greatest.

09
17
25, 27
30
44, 45, 46
56
61, 68
72, 77

Step 2

Use the digits in the tens place as the stems. Use the digits in the ones place as the leaves. Write leaves in increasing order.

Stem	Leaves
0	9
1	7
2	5 7
3	0
4	4 5 6
5	6
6	1 8
7	2 7

Ask questions such as these as you work together.

What reason might a person have for organizing data in a stem-and-leaf plot? Your child might reply: A stem-and-leaf plot shows each item of the data.

How would you organize data that has items in the hundreds? Your child might reason: I could make the stems using the digits in the hundreds and tens places and the leaves using the digits from the ones place.

Lower extreme The least value in the data

Upper extreme The greatest value in the data

Lower quartile The median of the lower half of the data

Upper quartile The median of the upper half of the data

If we added 79 and 58 to the data about countries, how would it change the plot? Your child might respond: The leaves for the 7 stem would be 2, 7, 9 and the 5 stem would have leaves of 6, 8.

This is how your child is learning to make and understand box-and-whisker graphs.

Value of Coins in Pockets

32 21 26 30 20 29 35 24
20 21 24 26 30 32

Step 1

Order the data.

20, 20, 21, 21, 24, 24, 26, 26, 29, 30, 30, 32, 32, 35

Step 2

Determine the following values:

- the median.
- the lower extreme, or the least value, and the upper extreme, or the greatest value.
- the lower quartile, or median of the lower half of the data, and the upper quartile, or the median of the upper half of the data.

Step 3

Create a box-and-whisker graph from the values.

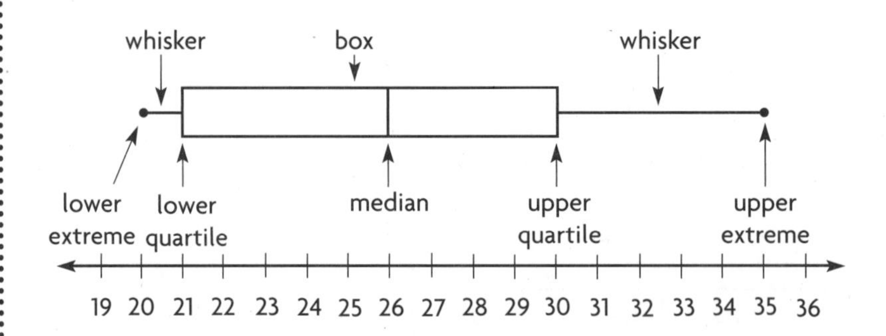

What measures of central tendency can you not find on a box-and-whisker graph? Your child might respond: A box-and-whisker graph does not show the mean or the mode.

As you work with your child, talk about math to help build confidence and understanding.

Sincerely,

Name ___________________________

Graph Data

Use the data in the table.

1. Make a multiple-bar graph.
2. Make a multiple-line graph.

Number of Sodas Consumed

	Friday	Saturday	Sunday
Boys	29	51	47
Girls	21	56	40

Use the data in the table.

3. Make a line graph.
4. If the trend continues, how many miles will be run on the sixth day? ______

Distance Run

Day	1	2	3	4	5
Miles	2	3.5	5	6.5	8

Make a stem-and-leaf plot of this data.

38	46	35	34	32	42	25	28	38	22	29	36

5. Find the mode of the data. ______
6. Find the median of the data. ______

Use the box-and-whisker graph. The data shows points scored by a baseball team during a tournament.

7. What was the greatest number of runs scored? ______
8. What was the fewest number of runs scored? ______
9. What is the median? ______
10. What are the lower and upper quartiles? ______

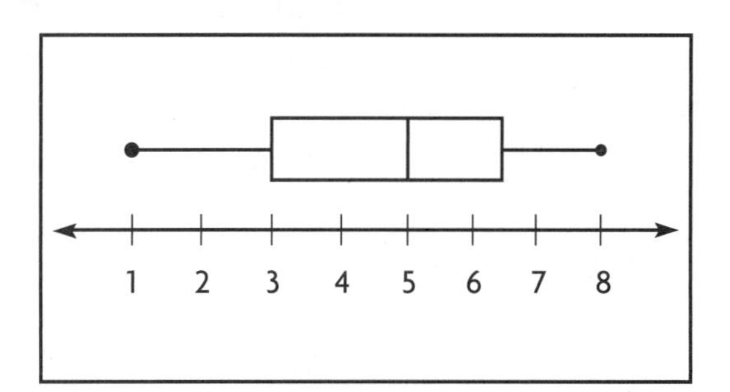

Use the bar graph.

11. Why is this graph misleading? ______________________
12. How could you change it so that it is not misleading?

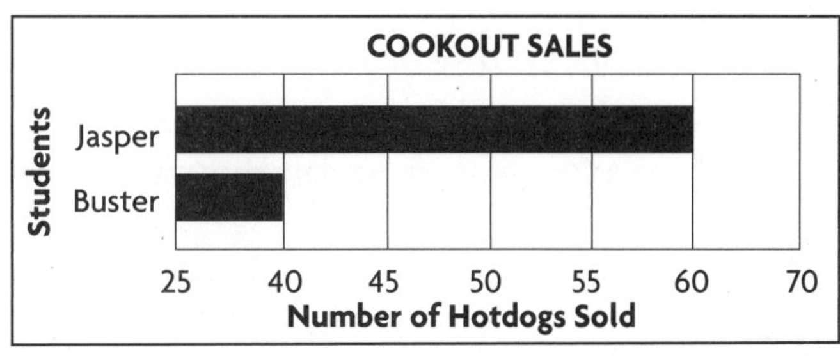

Answers: For 1-3, check students' graphs. 4. 9.5 miles; 5. 38; 6. 34.5; 7. 8; 8. 1; 9. 5; 10. 3 and 6.5; 11. It appears that Jasper sold more than four times as many hot dogs as Buster.; 12. Start the scale at zero and have equal intervals.

Family Fun WHOYAAH!

Materials

- 2 number cubes
- math fact charts
- a set of bonus cards with the following facts:
  7×1 through 7×9; 8×1 through 8×9; 9×1 through 9×9

Objective: Complete a row or column in a math-fact chart before your opponent. Players keep a tally of the number of cube tosses in each game.

Directions

1. Each player uses a new math fact chart for each game.
2. Each player tosses the number cubes and keeps a tally of the number of tosses in the table below.
3. The two numbers shown on the cubes are factors. The player shades the product in the correct square on the math fact chart. Except for doubles facts, each product may be recorded in one of two places on the chart. (If 3 and 5 are tossed, the square at the intersection of 3 across and 5 down or the square at the intersection of 5 across and 3 down could be shaded.)
4. If a player rolls a double and knows the product, he or she shades in the square and draws a bonus card, gives the correct product, and records that on the chart as well. The winner is the first player to complete a row or column.

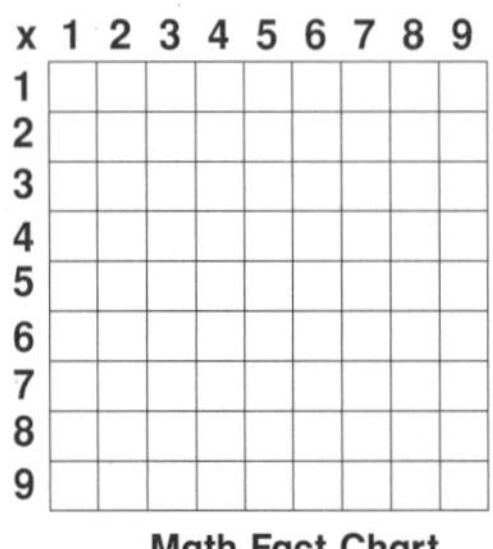

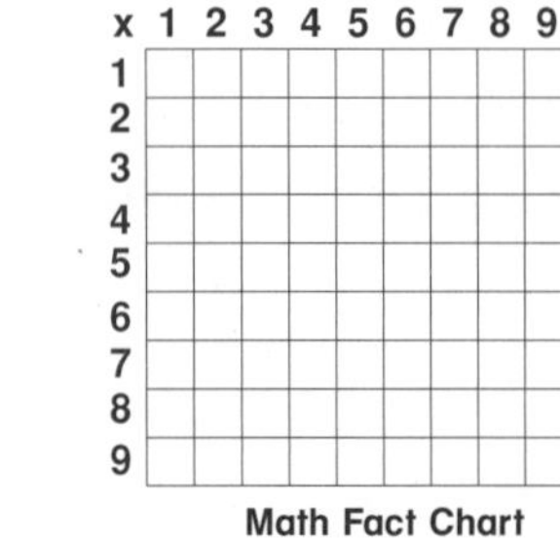

Math Fact Chart

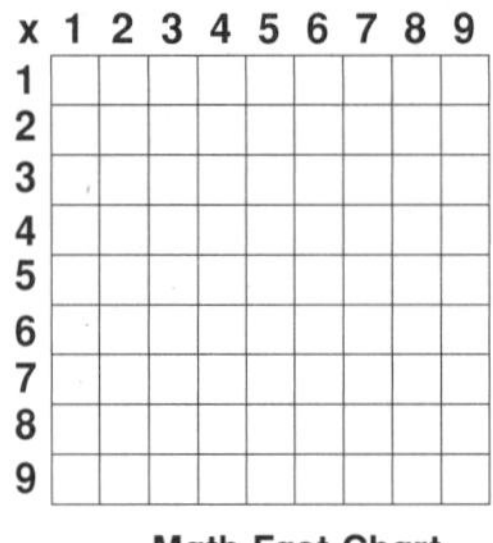

Math Fact Chart

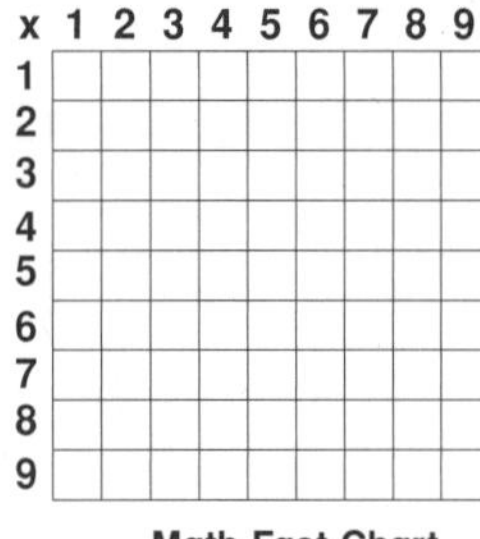

Math Fact Chart

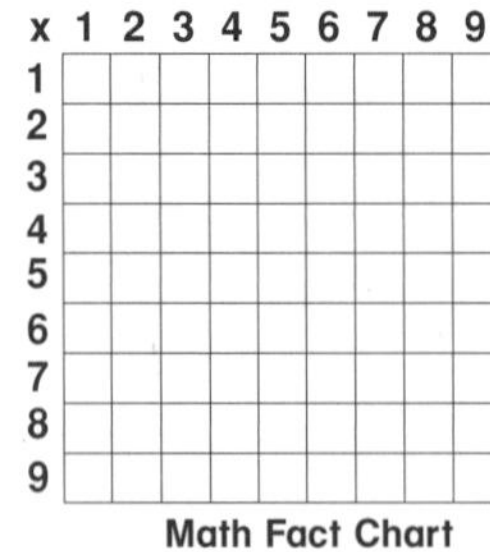

Math Fact Chart

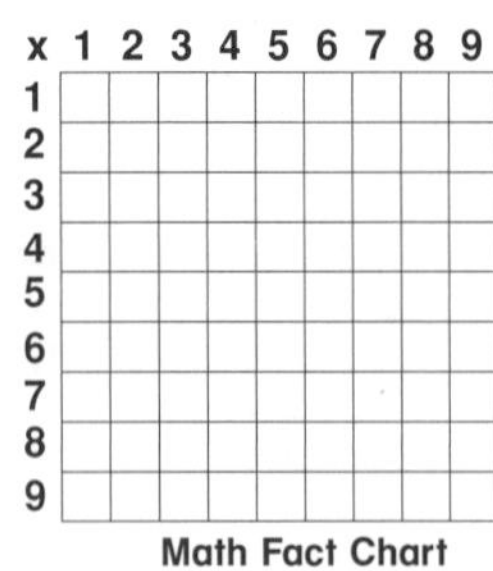

Math Fact Chart

After five games the players make a graph showing the number of throws per game.

Player	Game 1	Game 2	Game 3	Game 4	Game 5

Name

Date

Chapter 7

WHAT WE ARE LEARNING

Number Theory

VOCABULARY

Here are some of the vocabulary words we use in class:

Prime factorization A composite number expressed as the product of prime factors

Least common multiple (LCM) The smallest of the common multiples of two or more numbers

Greatest common factor (GCF) The largest of the common factors of two or more numbers

Dear Family,

Your child is working on divisibility rules, finding prime factors, and determining the least common multiple (LCM) and the greatest common factor (GCF) of whole numbers.

Find the prime factors of 120.

Step 1

Repeatedly divide by the smallest possible prime factor until the quotient is 1.

$\frac{120}{2} = 60$ $\frac{60}{2} = 30$ $\frac{30}{2} = 15$ $\frac{15}{3} = 5$ $\frac{5}{5} = 1$

Step 2

List the prime numbers you divided by. These are the prime factors.

2, 2, 2, 3, 5

So, the prime factorization of 120 is $2 \times 2 \times 2 \times 3 \times 5$ or $2^3 \times 3 \times 5$.

Use a factor tree to find the prime factors of 156.

Step 1

Choose any two factors of 156. Continue factoring until only prime factors are left.

156
4 39
2 2 3 13

So, the prime factorization of 156 is $2 \times 2 \times 3 \times 13$.

This is how your child is learning to find the least common multiple of numbers.

Find the LCM of 3 and 4.

Step 1	Step 2	Step 3
List multiples of each the number. 3: 3, 6, 9, 12, 15, 18, 21, 24, 4: 4, 8, 12, 16, 20, 24	Find the common multiples. 12, 24	Find LCM. 12

Ask questions such as these as you work together.

What is the first step in finding the LCM of 3 and 5? Your child might reply: First I list multiples of each number. 3: 3, 6, 9, 12, 15, 18, 21, 24, 27, 30 and 5: 5, 10, 15, 20, 25, 30.

What is your next step? Your child might explain:

Then I find the common multiples, 15 and 30.

What is the final step? Your child might respond: Since 15 is less than 30, 15 is the LCM.

This is how your child is learning to find the greatest common factor of numbers.

Find the GCF of 36 and 48.

Step 1	Step 2	Step 3
List all the factors of each number.	Find the common factors.	Identify the GCF.
36: 1, 2, 3, 4, 6, 12 48: 1, 2, 3, 4, 6, 8, 12	1, 2, 3, 4, 6, 12	12

As you work with your child, talk about math to help build confidence and understanding.

Sincerely,

Name ____________________

PRACTICE/HOMEWORK

Number Theory

Determine whether each number is divisible by 2, 3, 4, 5, 6, 8, 9, or 10.

1. 44 ________ 2. 300 ________ 3. 842 ________ 4. 2,645 ________

5. 6,009 ________ 6. 1,072 ________ 7. 459 ________ 8. 550 ________

9. 225 ________ 10. 15,000 ________

11. Without dividing, how would you know that 11,322 is divisible by 9?

____

12. Without dividing, how would you know that 32,644 is divisible by 4?

____

Find the LCM of each set of numbers.

13. 2, 5 ______

14. 3, 4 ______

15. 4, 6 ______

16. 4, 5, 10 ______

17. 6, 8, 12 ______

18. 6, 9, 12 ______

19. 4, 8, 10 ______

20. 25, 115 ______

21. 27, 12 ______

22. 24, 48 ______

Find the GCF of each set of numbers.

23. 48, 20 ______ 24. 72, 54 ______

25. 13, 24 ______ 26. 355, 300 ______

27. 12, 8, 20 ______ 28. 40, 24 ______

29. 303, 21 ______ 30. 27, 189 ______

31. 8, 16, 20 ______ 32. 33, 25 ______

Answers: **1.** 2, 4; **2.** 2, 3, 4, 5, 6, 10; **3.** 2; **4.** 5; **5.** 3; **6.** 2, 4, 8; **7.** 3, 9; **8.** 2, 5, 10; **9.** 3, 5, 9; **10.** 2, 3, 4, 5, 6, 8, 10; **11.** The digits add up to 9.; **12.** The last two digits are divisible by 4.; **13.** 10; **14.** 12; **15.** 12; **16.** 20; **17.** 24; **18.** 36; **19.** 40; **20.** 575; **21.** 108; **22.** 48; **23.** 4; **24.** 18; **25.** 1; **26.** 5; **27.** 4; **28.** 8; **29.** 3; **30.** 27; **31.** 4; **32.** 1

Family Fun

Color Squares

Directions:

- If a number is divisible by 3, color the square red.
- If a number is divisible by 4, color the square black.

15	16	153	56	60
48	99	64	621	100
276	28	3	52	368
92	243	444	726	40
324	624	531	76	810

Create your own design.

Name

GRADE 6

Chapter 8

Date

WHAT WE ARE LEARNING

Fraction Concepts

VOCABULARY

Here are some of the vocabulary words we use in class:

Equivalent fractions Fractions that name the same amount or the same part of a whole

Simplest form A fraction in which the numerator and denominator have no common factors other than 1

Mixed number A number that has a whole-number part and a fraction part

Terminating decimal A decimal that ends; a decimal for which the division operation results in a remainder of zero

Repeating decimal A decimal in which one or more digits repeat endlessly

Dear Family,

Your child is working on understanding fractions and mixed numbers, comparing and ordering fractions, and relating fractions, decimals, and percents.

This is how your child is learning to find equivalent fractions and write fractions in simplest form.

Find equivalent fractions by multiplying or dividing the numerator and the denominator by the same number.

Write a fraction equivalent to $\frac{2}{3}$.

$$\frac{2}{3} = \frac{2 \times 3}{3 \times 3} = \frac{6}{9}$$

Find the simplest form of a fraction by dividing by the greatest common factor.

Write the simplest form of $\frac{6}{12}$.

$$\frac{6}{12} = \frac{6 \div 6}{12 \div 6} = \frac{1}{2}$$

Find a mixed number by dividing the numerator by the denominator and using the quotient as the whole number part, the remainder as the numerator, and keeping the divisor as the denominator.

Write $\frac{17}{4}$ as a mixed number.

$$17 \div 4 = 4\frac{1}{4}$$

This is how your child is learning to write numbers as fractions, decimals, and percents

To write a decimal as a fraction, use the decimal place value to write the fraction.

$$0.9 = \frac{9}{10} \qquad 0.37 = \frac{37}{100}$$

To write a fraction as a decimal, divide the numerator by the denominator.

Write $\frac{3}{5}$ as a decimal.

$$3 \div 5 = 0.6$$

To write a fraction as a percent, convert the fraction to a decimal. Then write it as a percent.
Write $\frac{3}{5}$ as a percent.

$\frac{3}{5} = 0.6$

$0.6 = 60\%$

To compare fractions, decimals, and percents, write the expressions as decimals. Then compare decimals.
Compare 60%, $\frac{5}{8}$, and 0.37.

$60\% = 0.6$; $\frac{5}{8} = 0.625$

$0.625 > 0.6 > 0.37$, so $\frac{5}{8} > 60\% > 0.37$

As you work with your child, talk about math to help build confidence and understanding.

Sincerely,

Name ______________________

Fraction Concepts

Write each fraction in simplest form.

1. $\frac{15}{25}$ ___ 2. $\frac{6}{8}$ ___ 3. $\frac{6}{24}$ ___ 4. $\frac{24}{40}$ ___ 5. $\frac{21}{63}$ ___

6. $\frac{15}{20}$ ___ 7. $\frac{30}{36}$ ___ 8. $\frac{20}{28}$ ___ 9. $\frac{14}{42}$ ___ 10. $\frac{75}{150}$ ___

Write each fraction as a mixed number and each mixed number as a fraction in simplest form.

11. $\frac{8}{3}$ ___ 12. $\frac{72}{5}$ ___ 13. $\frac{17}{8}$ ___ 14. $\frac{63}{4}$ ___ 15. $\frac{12}{8}$ ___

16. $6\frac{1}{3}$ ___ 17. $9\frac{3}{5}$ ___ 18. $11\frac{3}{8}$ ___ 19. $19\frac{1}{2}$ ___ 20. $14\frac{1}{3}$ ___

Write each decimal as a fraction and each fraction as a decimal.

21. 0.31 ___ 22. $\frac{2}{3}$ ___ 23. 0.617 ___ 24. $\frac{7}{28}$ ___ 25. 0.4 ___

26. $\frac{3}{8}$ ___ 27. 0.09 ___ 28. $\frac{3}{20}$ ___ 29. 0.85 ___ 30. $\frac{23}{50}$ ___

Write each fraction as a decimal and a percent.

31. $\frac{3}{4}$ ____ 32. $\frac{5}{8}$ ____ 33. $\frac{4}{5}$ ____ 34. $\frac{6}{25}$ ____ 35. $\frac{9}{10}$ ____

Compare the decimals and fractions. Write <, > or = for each circle.

36. 0.1 ◯ $\frac{1}{6}$ 37. $\frac{11}{14}$ ◯ 0.7 38. $\frac{5}{6}$ ◯ 0.75

39. 0.03 ◯ $\frac{1}{50}$ 40. $\frac{34}{3}$ ◯ $\frac{50}{4}$ 41. 0.6 ◯ 0.09

Answers: **1.** $\frac{3}{5}$; **2.** $\frac{3}{4}$; **3.** $\frac{1}{4}$; **4.** $\frac{3}{5}$; **5.** $\frac{1}{3}$; **6.** $\frac{3}{4}$; **7.** $\frac{5}{6}$; **8.** $\frac{5}{7}$; **9.** $\frac{1}{3}$; **10.** $\frac{1}{2}$; **11.** $2\frac{2}{3}$; **12.** $14\frac{2}{5}$; **13.** $2\frac{1}{8}$; **14.** $15\frac{3}{4}$; **15.** $1\frac{1}{2}$; **16.** $\frac{19}{3}$; **17.** $\frac{48}{5}$; **18.** $\frac{91}{8}$; **19.** $\frac{39}{2}$; **20.** $\frac{43}{3}$; **21.** $\frac{31}{100}$; **22.** $0.\overline{6}$; **23.** $\frac{617}{1000}$; **24.** 0.25; **25.** $\frac{4}{10}$, or $\frac{2}{5}$; **26.** 0.375; **27.** $\frac{9}{100}$; **28.** 0.15; **29.** $\frac{85}{100}$, or $\frac{17}{20}$; **30.** 0.46; **31.** 0.75, 75%; **32.** 0.625, 62.5%; **33.** 0.8, 80%; **34.** 0.24, 24%; **35.** 0.9, 90%; **36.** <; **37.** >; **38.** >; **39.** >; **40.** <; **41.** >

Family Fun MATCHING GAME

MATH GAME

Materials

- 30 fraction/decimal/percent cards

Directions

1. Cut out the cards.
2. Shuffle the cards and put them face down in a 6 by 7 array.
3. Players take turns turning over 2 cards.
4. If the cards are equivalent, the player keeps them. If the cards are not equivalent, the player turns the cards back over in the same place.
5. The game is over when all of the cards have been matched.
6. The player with the most cards wins.

0.1	0.5	0.25	0.2	0.3	0.4	0.75
0.7	0.9	10%	50%	$33\frac{1}{3}\%$	25%	20%
30%	40%	$66\frac{2}{3}\%$	75%	70%	90%	$\frac{1}{10}$
$\frac{1}{2}$	$\frac{1}{3}$	$\frac{1}{4}$	$\frac{1}{5}$	$\frac{3}{10}$	$\frac{2}{5}$	$\frac{3}{5}$
$\frac{2}{3}$	$\frac{3}{4}$	$\frac{7}{10}$	$\frac{9}{10}$	$\frac{2}{20}$	$\frac{5}{10}$	$\frac{2}{8}$
$\frac{2}{10}$	$\frac{6}{20}$	$\frac{4}{10}$	$\frac{6}{10}$	$\frac{6}{8}$	$\frac{14}{20}$	$\frac{18}{20}$

Name

Date

Chapter 9

WHAT WE ARE LEARNING

Add and Subtract Fractions and Mixed Numbers

VOCABULARY

Here is a vocabulary term we use in class:

Unlike fractions Fractions with different denominators

Dear Family,

Your child is continuing the study of fractions by learning how to add and subtract with fractions and mixed numbers.

This is how your child is learning to estimate sums and differences of fractions and mixed numbers.

To estimate with fractions less than 1, compare the numerator to the denominator to decide if the fraction is close to 0, $\frac{1}{2}$, or 1.

Estimate the sum of $\frac{7}{10} + \frac{3}{8}$.

$\frac{7}{10} + \frac{3}{8}$ — $\frac{7}{10}$ is closer to $\frac{1}{2}$ than 1.

$\frac{3}{8}$ is closer to $\frac{1}{2}$ than 0.

$\frac{1}{2} + \frac{1}{2} = 1$; so, $\frac{7}{10} + \frac{3}{8}$ is about 1.

To estimate sums and differences of mixed numbers, round each mixed number to the nearest whole number.

Estimate the sum of $2\frac{1}{3} + 1\frac{1}{6}$.

$2\frac{1}{3} + 1\frac{1}{6}$ — $2\frac{1}{3}$ is closer to 2 than 3.

$1\frac{1}{6}$ is closer to 1 than 2.

$2 + 1 = 3$; so, $2\frac{1}{3} + 1\frac{1}{6}$ is about 3.

To estimate sums and differences in another way, find the range using two estimates.

$6\frac{3}{4} - 3\frac{7}{8}$

$6\frac{3}{4}$ is close to 7.

$3\frac{7}{8}$ is close to 4.

$7 - 4 = 3$

$6\frac{3}{4} - 3\frac{7}{8}$

$6\frac{3}{4}$ is close to $6\frac{1}{2}$.

$3\frac{7}{8}$ is close to 4.

$6\frac{1}{2} - 4 = 2\frac{1}{2}$

The range for the difference is $2\frac{1}{2}$ to 3. So a good estimate of $6\frac{3}{4} - 3\frac{7}{8}$ would be $2\frac{3}{4}$ because that is halfway between $2\frac{1}{2}$ and 3.

This is how your child is learning to add and subtract fractions and mixed fractions.

Find the difference. $5\frac{1}{4} - 3\frac{2}{3}$.

Step 1

Estimate.

$5\frac{1}{4}$ is close to 5.

$3\frac{2}{3}$ is close to $3\frac{1}{2}$.

So, the difference is about $5 - 3\frac{1}{2}$, or $1\frac{1}{2}$.

Step 2

Write equivalent fractions by using the least common denominator.

$$\begin{array}{r} 5\frac{1}{4} = 5\frac{3}{12} \\ -3\frac{2}{3} = 3\frac{8}{12} \\ \hline \end{array}$$

Step 3

Since $\frac{8}{12}$ is greater than $\frac{3}{12}$, rename $5\frac{3}{12}$ as $4\frac{15}{12}$.

$$\begin{array}{r} 5\frac{1}{4} = 5\frac{3}{12} = 4\frac{15}{12} \\ -3\frac{2}{3} = 3\frac{8}{12} = 3\frac{8}{12} \\ \hline \end{array}$$

Step 4

Subtract.

$$\begin{array}{r} 5\frac{1}{4} = 5\frac{3}{12} = 4\frac{15}{12} \\ -3\frac{2}{3} = 3\frac{8}{12} = 3\frac{8}{12} \\ \hline 1\frac{7}{12} \end{array}$$

The answer is reasonable because it is close to the estimate of $1\frac{1}{2}$.

Ask your child how to add $7\frac{1}{8}$ and $1\frac{1}{6}$. Your child might suggest: First I would make equivalent fractions by using the LCD. $7\frac{1}{8} = 7\frac{3}{24}$ and $1\frac{1}{6} = 1\frac{4}{24}$.

What is the next step? Your child might respond: I add the fractions by adding the numerators and writing the sum over the denominator. Then I add the whole numbers. $8\frac{7}{24}$

How can you estimate to check your answer? Your child might reason: $7\frac{1}{8}$ is close to 7 and $1\frac{1}{6}$ is close to 1, so my estimate is 8. That is close to the actual answer.

As you work with your child, talk about math to help build confidence and understanding.

Sincerely,

Name ______________________

PRACTICE/HOMEWORK

Add and Subtract Fractions and Mixed Numbers

Estimate the sum or difference.

1. $\frac{5}{8} - \frac{5}{9}$ ______
2. $5\frac{1}{3} + 2\frac{4}{5}$ ______
3. $\frac{9}{10} + 4$ ______
4. $5\frac{7}{8} - 2\frac{1}{16}$ ______
5. $\frac{1}{4} - \frac{1}{5}$ ______
6. $7\frac{4}{7} + 4\frac{1}{3}$ ______
7. $1\frac{1}{10} - \frac{2}{25}$ ______
8. $9\frac{2}{7} + \frac{1}{3}$ ______
9. $9\frac{7}{8} - 3\frac{1}{3}$ ______

Write the sum or the difference in simplest form.

10. $\frac{1}{3} + \frac{1}{6}$ ______
11. $\frac{5}{8} - \frac{1}{16}$ ______
12. $\frac{7}{12} - \frac{1}{6}$ ______
13. $\frac{1}{3} - \frac{1}{7}$ ______
14. $\frac{3}{14} - \frac{1}{7}$ ______
15. $\frac{1}{5} + 4\frac{9}{10}$ ______
16. $\frac{5}{9} - \frac{1}{3}$ ______
17. $\frac{8}{9} - \frac{1}{3}$ ______
18. $\frac{3}{4} + \frac{1}{10}$ ______

Draw a diagram to find the sum or difference. Write the answer in simplest form.

19. $5\frac{5}{12} - 1\frac{1}{6}$
20. $3\frac{1}{5} + 1\frac{9}{10}$
21. $5\frac{1}{3} - 3\frac{1}{9}$
22. $4\frac{1}{4} - 2\frac{5}{8}$

Diagram

Answer

Answers: 1. 0; 2. 8; 3. 5; 4. 4; 5. 0; 6. 12; 7. 1; 8. $9\frac{1}{2}$; 9. 7; 10. $\frac{1}{2}$; 11. $\frac{9}{16}$; 12. $\frac{5}{12}$; 13. $\frac{4}{21}$; 14. $\frac{1}{14}$; 15. $5\frac{1}{10}$; 16. $\frac{2}{9}$; 17. $\frac{5}{9}$; 18. $\frac{17}{20}$;
For 19-22, check students' diagrams. 19. $4\frac{1}{4}$; 20. $5\frac{1}{10}$; 21. $2\frac{2}{9}$; 22. $1\frac{5}{8}$

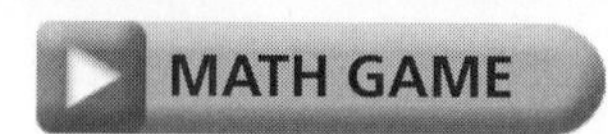

Family Fun

An animal found on an island in the Pacific Ocean can live for a year without food or water. It can live more than 150 years altogether. Yet despite these amazing survival abilities, the animal is an endangered species facing extinction. To find the name of the animal, solve the puzzle.

Directions

- Work each exercise.
- Locate the whole number part of your answer in the column to the left of the letter box. Locate the fractional part of your answer in the row at the top. Find the letter at the intersection of the row and column.
- Write the letter above the number of the exercise at the bottom of the page.

1. $\frac{3}{4} + \frac{3}{4}$
2. $\frac{7}{8} + \frac{3}{8}$
3. $2\frac{5}{6} - \frac{1}{2}$
4. $3 - 1\frac{3}{4}$
5. $1\frac{1}{3} + 1\frac{1}{2}$
6. $2\frac{7}{8} - 1\frac{5}{8}$
7. $\frac{7}{10} + \frac{4}{5}$
8. $3\frac{5}{12} - \frac{11}{12}$
9. $2\frac{1}{2} + \frac{5}{6}$
10. $1\frac{7}{8} + 1\frac{1}{2}$
11. $4\frac{1}{6} - 1\frac{2}{3}$
12. $\frac{11}{12} + 2\frac{1}{3}$
13. $5\frac{5}{24} - 1\frac{5}{6}$
14. $\frac{9}{10} + 1\frac{3}{5}$
15. $6\frac{1}{4} - 4$
16. $5\frac{1}{12} - 1\frac{3}{4}$
17. $\frac{5}{8} + \frac{3}{4}$

	$\frac{1}{4}$	$\frac{1}{3}$	$\frac{3}{8}$	$\frac{2}{5}$	$\frac{1}{2}$	$\frac{5}{6}$
1	A	C	E	F	G	H
2	I	L	M	N	O	P
3	R	S	T	U	V	W

Answer: ____ 1 ____ 2 ____ 3 ____ 4 ____ 5 ____ 6 ____ 7 ____ 8 ____ 9

____ 10 ____ 11 ____ 12 ____ 13 ____ 14 ____ 15 ____ 16 ____ 17

Answer: Galapagos tortoise

HARCOUNT MATH

GRADE 6

Chapter 10

WHAT WE ARE LEARNING

Multiplying and Dividing Fractions and Mixed Numbers

VOCABULARY

Here is a vocabulary term we use in class:

Reciprocal One of two numbers whose product is 1

$\frac{3}{5} \times \frac{5}{3} = 1$

$\frac{3}{5}$ and $\frac{5}{3}$ are reciprocals.

Name

Date

Dear Family,

Your child is continuing the study of fractions.

This is how your child is learning to estimate products and quotients.

Step 1

Round mixed numbers to the nearest whole number and fractions to 1, $\frac{1}{2}$, or 0, whichever is nearest.

$48\frac{7}{10} \times 5\frac{2}{5}$ $\quad$ 49×5

$48\frac{7}{10} \div 5\frac{2}{5}$ $\quad$ $49 \div 5$

Step 2

Perform the operation.

$49 \times 5 = 245$ $\quad$ $49 \div 5 = 9\frac{4}{5}$

How would you estimate the product $9\frac{1}{8} \times \frac{3}{4}$? Your child might suggest: The first thing to do is round the numbers. $9\frac{1}{8}$ is close to 9 and $\frac{3}{4}$ is close to 1.

What do you do next? Your child might respond: I perform the operation in my head.

$9 \times 1 = 9$

How would you estimate the quotient $9\frac{1}{8} \div \frac{1}{3}$?

$9\frac{1}{8}$ is close to 9 and $\frac{1}{3}$ is close to $\frac{1}{2}$.

$9 \div \frac{1}{2} = 18$

Why is your estimate for the quotient of $9\frac{1}{8} \div \frac{1}{3}$ so much greater than the dividend? One way to think about $9\frac{1}{8} \div \frac{1}{3}$ is "how many $\frac{1}{3}$ pieces are in $9\frac{1}{8}$ pieces?"

The answer is greater than $9\frac{1}{8}$.

This is how your child is learning to multiply and divide fractions and mixed numbers.

Step 1

Write mixed numbers as fractions. If dividing, use the reciprocal of the divisor to write a multiplication expression.

$1\frac{1}{3} \times 1\frac{1}{5} = \frac{4}{3} \times \frac{6}{5}$

$1\frac{1}{3} \div 1\frac{1}{5} = \frac{4}{3} \div \frac{6}{5} = \frac{4}{3} \times \frac{5}{6}$

Step 2

Simplify and multiply the numerators and the denominators.

$\frac{4 \times 6}{3 \times 5} = \frac{8}{5}$, or $1\frac{3}{5}$

$\frac{4 \times 5}{3 \times 6} = \frac{10}{9}$, or $1\frac{1}{9}$

As you work with your child, talk about math to help build confidence and understanding.

Sincerely,

Name ______________________

PRACTICE/HOMEWORK

Multiply and Divide Fractions and Mixed Numbers

Estimate the products and quotients.

1. $\frac{5}{9} \times \frac{2}{5}$ ____ 2. $8\frac{3}{7} \div 4$ ____ 3. $8\frac{1}{12} \div 3\frac{2}{3}$ ____

4. $5\frac{9}{22} \times 7\frac{4}{7}$ ____ 5. $4\frac{9}{10} \div \frac{5}{9}$ ____ 6. $77\frac{2}{7} \div 23\frac{1}{3}$ ____

7. $15\frac{5}{8} \div 4\frac{1}{9}$ ____ 8. $120\frac{1}{3} \div 11\frac{19}{20}$ ____ 9. $\frac{2}{3} \div \frac{3}{4}$ ____

Find the product. Write the answer in simplest form.

10. $\frac{2}{3} \times \frac{1}{5}$ ____ 11. $16 \times \frac{7}{12}$ ____ 12. $\frac{2}{3} \times 1$ ____

13. $1\frac{1}{3} \times 4$ ____ 14. $3\frac{1}{3} \times 2\frac{2}{5}$ ____ 15. $1\frac{1}{5} \times 3$ ____

Find the quotient. Write the answer in simplest form.

16. $\frac{1}{4} \div \frac{2}{3}$ ____ 17. $3\frac{1}{3} \div 2\frac{2}{5}$ ____ 18. $4 \div \frac{6}{7}$ ____

19. $\frac{3}{8} \div 3$ ____ 20. $12 \div \frac{3}{5}$ ____ 21. $1\frac{3}{5} \div 4$ ____

Evaluate the expressions.

22. $2x,\ x = 5\frac{1}{5}$ ____ 23. $\frac{1}{4}y,\ y =$ ____ 24. $c \div 3\frac{3}{4},\ c = 2\frac{3}{8}$ ____

Solve the equations.

25. $x + 1\frac{1}{3} = 6\frac{2}{3}$ ____ 26. $x - 3\frac{7}{8} = 4\frac{3}{16}$ ____ 27. $22\frac{1}{4} + x = 37\frac{7}{8}$ ____

Answers: For 1-9, possible estimates are given. **1.** $\frac{1}{2}$; **2.** 2; **3.** 2; **4.** 40; **5.** 10; **6.** 3; **7.** 4; **8.** 10; **9.** 1; **10.** $\frac{2}{15}$; 11. $9\frac{1}{3}$; **12.** 1; **13.** $6\frac{1}{3}$; **14.** 8; **15.** $4\frac{1}{2}$; **16.** $\frac{3}{8}$; **17.** $1\frac{7}{18}$; **18.** $4\frac{2}{3}$; **19.** $\frac{1}{8}$; **20.** 20; **21.** $\frac{2}{5}$; **22.** $10\frac{2}{5}$; **23.** $\frac{1}{8}$; **24.** $\frac{19}{30}$; **25.** $5\frac{1}{3}$; **26.** $8\frac{1}{16}$; **27.** $15\frac{5}{8}$

Family Fun

A Fraction Mathematicalosterm

MATH GAME

A mathematicalosterm is like a word search except the combinations are not words but math problems and answers. On this puzzle there are at least ten problems that involve multiplying or dividing fractions and mixed numbers. Number combinations appear horizontally, vertically, and diagonally. They can be backward or forward.

Directions: Circle the three numbers that complete a combination correctly. Then record the problem with the correct operation sign. One combination is circled for you. Why is it circled? Because $\frac{1}{3} \div \frac{1}{6}$ equals 2!

$\frac{1}{3}$	$\frac{1}{6}$	2	$\frac{1}{4}$	$\frac{1}{2}$	$\frac{1}{2}$
$\frac{3}{4}$	$3\frac{2}{3}$	$1\frac{1}{6}$	$\frac{5}{8}$	$3\frac{1}{2}$	$2\frac{1}{2}$
$\frac{1}{4}$	$\frac{3}{8}$	$\frac{3}{10}$	$1\frac{3}{4}$	1	$\frac{2}{5}$
$2\frac{1}{3}$	$1\frac{3}{8}$	$6\frac{2}{3}$	$3\frac{1}{4}$	$\frac{2}{3}$	1
$3\frac{1}{3}$	4	$1\frac{2}{3}$	$2\frac{1}{2}$	$\frac{2}{3}$	$2\frac{7}{8}$
$\frac{5}{9}$	$1\frac{3}{10}$	$\frac{4}{5}$	$1\frac{5}{8}$	$\frac{4}{9}$	$2\frac{3}{4}$

HARCOURT MATH

GRADE 6

Chapter 11

WHAT WE ARE LEARNING

Number Relationships

Vocabulary

Here are some of the vocabulary words we use in class:

Integers All whole numbers and their opposites

Opposites Two integers that are the same distance from 0 but on the opposite side of 0

Positive integers Integers greater than 0

Negative integers Integers less than 0

Absolute value The distance of an integer from zero

Ratio A comparison of two numbers, a and b, written as a fraction $\frac{a}{b}$

Rational number Any number that can be written as a ratio $\frac{a}{b}$

Name

Date

Dear Family,

Your child is beginning the study of rational numbers.

This is how your child is learning to identify integers.

Integers are all whole numbers and their opposites. Positive integers, integers greater than 0, are written with or without a + sign. Negative integers, integers less than 0, are written with a – sign.

You can find the absolute value, or distance from 0, on a number line.

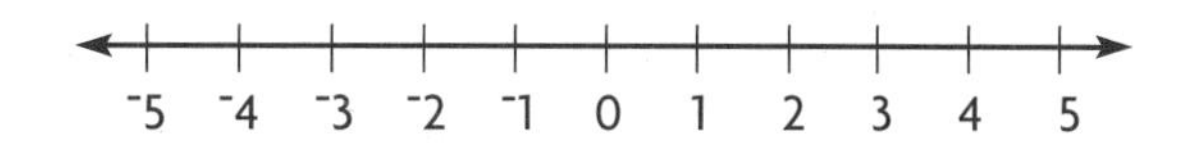

What is the absolute value of ⁻5? Your child might say 5 since ⁻5 is 5 units from 0.

What is an integer? Your child might respond: An integer is any whole number. It can be positive or negative. Zero is also an integer.

Name an integer that you might use to describe these situations: a temperature of 23 degrees below zero, a five point rise in your math grade, a 10-yard penalty in football. Your child might suggest: ⁻23°, ⁺5, ⁻10.

How do you write absolute values and how do you read them? Your child might explain: I write an absolute value like this: $|{}^-5| = 5$. I say, "The absolute value of negative five is five."

Venn diagram A diagram that shows how sets of numbers are related

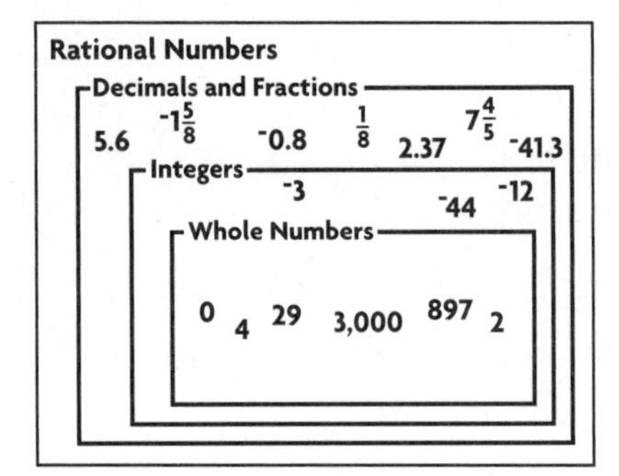

This is how your child is learning to classify rational numbers and to find a rational number between two rational numbers.

A rational number is any number that can be written as a ratio $\frac{a}{b}$. These are all rational numbers.

4 $\quad \frac{7}{10} \quad$ 56 $\quad$ $^-12.5$

To find a rational number between two rational numbers, you can use equivalent fractions or decimals.

Find a number between 5.5 and 6.

Write equivalent fractions or decimals.

5.5 = 5.50 $\qquad$ 6 = 6.00

5.75 is between 5.50 and 6.00, so 5.75 is one number between 5.5 and 6.

To compare rational numbers, write them as decimals or as fractions with a common denominator.

Compare $\frac{3}{4}$ and 0.7.

$\frac{3}{4} = 0.75$ and $0.7 = 0.70$

$0.75 > 0.70$, so $\frac{3}{4} > 0.7$.

As you work with your child, talk about math to help build confidence and understanding.

Sincerely,

Name ________________________

Number Relationships

Write the opposite integer.

1. $^-7$ _____ 2. 13 _____ 3. $^-91$ _____ 4. $^-782$ _____ 5. 50,423 _____

Write each rational number in the form $\frac{a}{b}$.

6. $4\frac{1}{7}$ _____ 7. 0.07 _____ 8. 22.3 _____ 9. $10\frac{1}{8}$ _____ 10. 43.9 _____

11. $3\frac{4}{9}$ _____ 12. 120 _____ 13. 0.81 _____ 14. $11\frac{3}{5}$ _____ 15. 1,903 _____

Find a rational number between the two given numbers.

16. $\frac{1}{8}$ and $\frac{1}{5}$ _____ 17. $\frac{5}{9}$ and $\frac{5}{8}$ _____ 18. $^-1.7$ and $^-1.5$ _____

19. 1.68 and 1.685 _____ 20. $\frac{-1}{4}$ and $\frac{-1}{8}$ _____ 21. $^-5.5$ and $^-5.3$ _____

22. 5.7 and 5.72 _____ 23. 6.89 and 6.9 _____ 24. $^-89.1$ and $^-89$ _____

Compare the rational numbers. Write <, >, or = for each ○.

25. 0.5 ○ 0.43 26. $\frac{3}{7}$ ○ 0.45 27. $^-0.70$ ○ $\frac{-3}{5}$

28. $^-0.04$ ○ $\frac{-1}{25}$ 29. 5.734 ○ $5\frac{5}{8}$ 30. $\frac{-4}{3}$ ○ $\frac{4}{3}$

Compare the rational numbers and order them from least to greatest.

31. $\frac{1}{2}$, $\frac{2}{5}$, 0.32, $\frac{1}{4}$ ____________

32. $^-0.3$, 0.03, 30, $^-30$ ____________

33. $\frac{1}{4}$, $\frac{1}{7}$, $\frac{1}{9}$, 0.1 ____________

Answers: 1. 7; **2.** $^-13$; **3.** 91; **4.** 782; **5.** $^-50,423$; **6.** $\frac{29}{7}$; **7.** $\frac{7}{100}$; **8.** $\frac{223}{10}$; **9.** $\frac{81}{8}$; **10.** $\frac{439}{10}$; **11.** $\frac{31}{9}$; **12.** $\frac{120}{1}$; **13.** $\frac{81}{100}$; **14.** $\frac{58}{5}$; **15.** $\frac{1903}{1}$;
For **16-24**, possible answers are given. **16.** $\frac{7}{40}$; **17.** $\frac{42}{72}$; **18.** $^-1.6$; **19.** 1.6825; **20.** $^-\frac{3}{16}$; **21.** $^-5.4$; **22.** 5.71; **23.** 6.895; **24.** $^-89.05$;
25. >; **26.** <; **27.** <; **28.** =; **29.** >; **30.** <; **31.** $\frac{1}{4}$, 0.32, $\frac{2}{5}$, $\frac{1}{2}$; **32.** $^-30$, $^-0.3$, 0.03, 30; **33.** 0.1, $\frac{1}{9}$, $\frac{1}{7}$, $\frac{1}{4}$

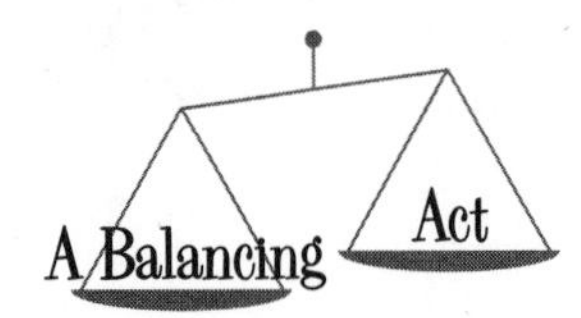

Picture the diagram below as a mobile that you might hang from a ceiling. At each balance point (•), the total value to the left must equal the total value to the right. The challenge is to determine the value of each of the geometric shapes.

The total value of the entire mobile is 24.

What are the values of the shapes? (Hint: Find the value of each triangle. Then work your way left.)

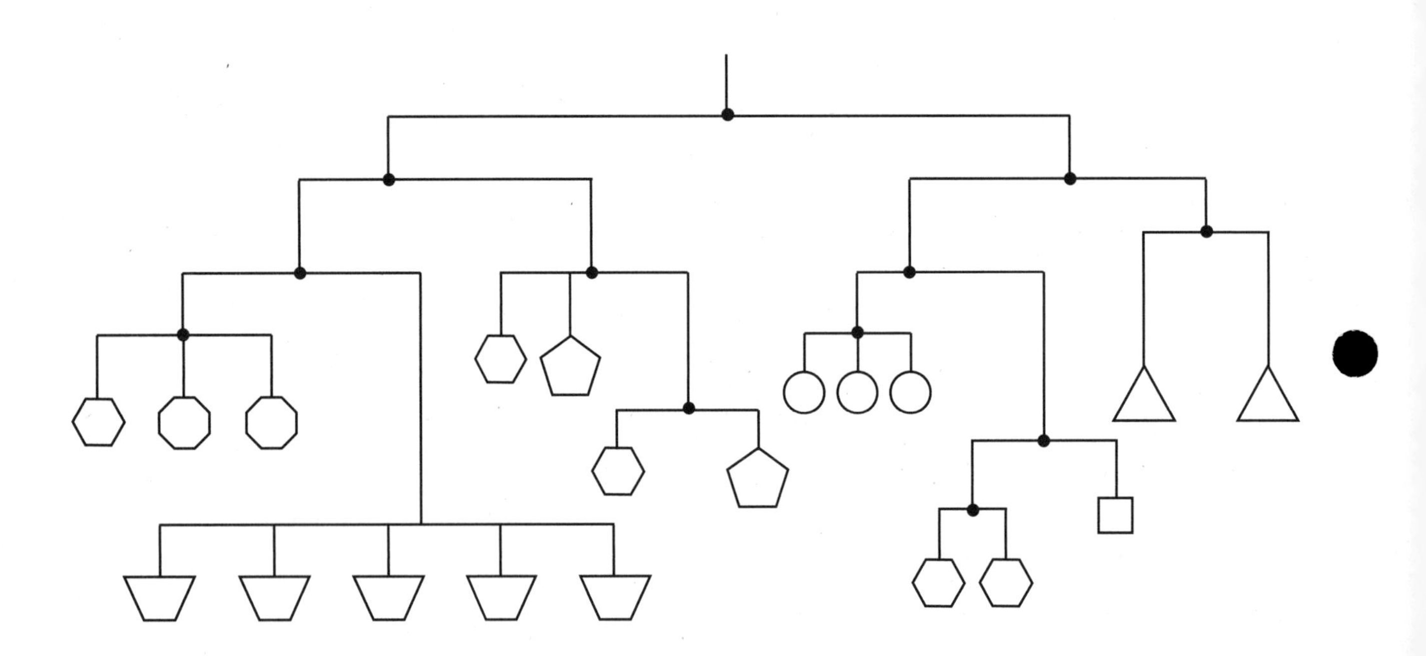

Answers: triangle = 3, circle = 1, square = $1\frac{1}{2}$, hexagon = $\frac{3}{4}$, pentagon = $2\frac{1}{4}$, octagon = $1\frac{1}{8}$, trapezoid = $\frac{3}{5}$

HARCOURT MATH

GRADE 6

Chapter 12

WHAT WE ARE LEARNING

Operations with Integers

Vocabulary

Here is one of the vocabulary words we use in class:

Additive inverse An integer with the opposite sign of another integer

Name

Date

Dear Family,

Your child is studying addition, subtraction, multiplication, and division with integers. Your child is learning about the rules for operations with integers by working with colored counters and also by using a number line. Your child will discover the rules for integers listed below and then use them to operate with rational numbers.

Rules for addition of integers

- To find the sum of two integers with like signs, add the absolute values of the integers and use the sign of the addends for the result.

  $4 + 3 = 7$ $\quad$ $^-4 + {}^-3 = {}^-7$

- To find the sum of two integers with unlike signs, subtract the lesser absolute value from the greater absolute value. Use the sign of the integer with greater absolute value for the result.

  $4 + {}^-3 = 1$ $\quad$ $^-4 + 3 = {}^-1$

Rules for subtraction of integers

- To find the difference of two integers with like signs, write the expression as an addition problem and apply the rules for addition of integers.

  $^-12 - {}^-3 = {}^-12 + (3) = {}^-9$

  $12 - 3 = 12 + ({}^-3) = 9$

- To find the difference of two integers with unlike signs, write the expression as an addition problem and apply the rules for addition of integers.

  $^-12 - (3) = {}^-12 + ({}^-3) = {}^-15$

  $12 - ({}^-3) = 12 + 3 = 15$

Rules for multiplication of integers

- The product of two integers with like signs is positive.

  $4 \times 3 = 12$ $\quad$ $^-4 \times {}^-3 = 12$

- The product of two integers with unlike signs is negative.

  $4 \times {}^-3 = {}^-12$ $\quad$ $^-4 \times 3 = {}^-12$

Rules for division of integers

- The quotient of two integers with like signs is positive.

  $\frac{^-12}{^-3} = 4$ $\frac{12}{3} = 4$

- The quotient of two integers with unlike signs is negative.

  $\frac{12}{^-3} = {^-4}$ $\frac{^-12}{3} = {^-4}$

Your child is learning to use the same rules to operate with rational numbers as were used when operating with integers.

$\frac{^-3}{4} \times \frac{^-5}{7} = \frac{15}{28}$ The fractions have like signs, so the product is positive.

Ask questions such as this as you work together.

Before we begin to work together with integers, can you tell me why I might say the rules for mulitplying and dividing integers are the same? Your child might explain: When I multiply or divide, if both signs are the same, the answer will always be positive. If the signs are different, the answer will always be negative.

As you work with your child, talk about math to help build confidence and understanding.

Sincerely,

Name ______________________

PRACTICE/HOMEWORK

Operations with Integers

Write the addition problem modeled on the number line.

1.

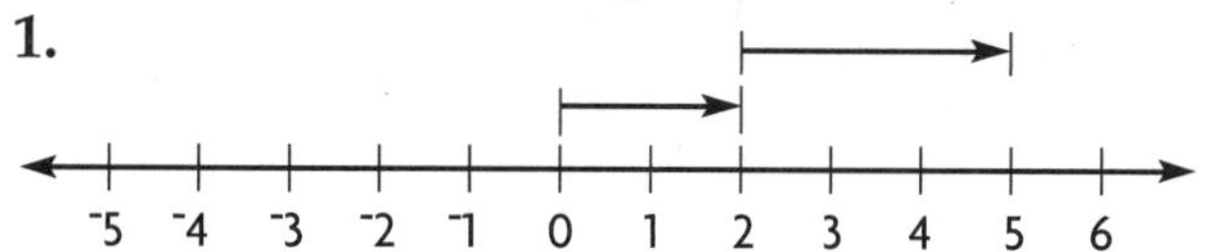

Add.

2. $^-6 + 19$ ____ 3. $^-8 + {}^-7$ ____ 4. $23 + {}^-19$ ____ 5. $^-15 + {}^-15$ ____

6. $52 + {}^-7$ ____ 7. $^-23 + {}^-23$ ____ 8. $^-18 + 33$ ____ 9. $^-15 + 2$ ____

Subtract.

10. $^-8 - {}^-2$ ____ 11. $4 - 9$ ____ 12. $5 - {}^-5$ ____ 13. $41 - {}^-23$ ____

14. $23 - {}^-15$ ____ 15. $^-43 - {}^-5$ ____ 16. $^-72 - 40$ ____ 17. $^-23 - {}^-15$ ____

Multiply.

18. $^-4 \times 8$ ____ 19. 6×11 ____ 20. $^-15 \times 3$ ____ 21. $60 \times {}^-7$ ____

22. $^-12 \times {}^-6$ ____ 23. $12 \times {}^-5$ ____ 24. $^-8 \times 15$ ____ 25. $^-2 \times {}^-16$ ____

Divide.

26. $^-10 \div {}^-2$ ____ 27. $40 \div 8$ ____ 28. $99 \div 9$ ____ 29. $60 \div 15$ ____

30. $^-96 \div 12$ ____ 31. $^-64 \div {}^-8$ ____ 32. $45 \div {}^-5$ ____ 33. $42 \div {}^-7$ ____

Add, subtract, multiply, or divide.

34. $^-4\frac{1}{2} - {}^-9$ ____ 35. $0.4 \times {}^-3.5$ ____ 36. $3.7 + {}^-11.4$ ____

37. $\frac{3}{4} \times \frac{4}{9}$ ____ 38. $^-10 \div {}^-2\frac{1}{2}$ ____ 39. $28.2 - {}^-9.7$ ____

Answers: **1.** 2 + 3 = 5; **2.** 13; **3.** ⁻15; **4.** 4; **5.** ⁻30; **6.** 45; **7.** ⁻46; **8.** 15; **9.** ⁻13; **10.** ⁻6; **11.** ⁻5; **12.** 10; **13.** 64; **14.** 38; **15.** ⁻38; **16.** ⁻112; **17.** ⁻8; **18.** ⁻32; **19.** 66; **20.** ⁻45; **21.** ⁻420; **22.** 72; **23.** ⁻60; **24.** ⁻120; **25.** 32; **26.** 5; **27.** 5; **28.** 11; **29.** 4; **30.** ⁻8; **31.** 8; **32.** ⁻9; **33.** ⁻6; **34.** 4.5; **35.** ⁻1.4; **36.** ⁻7.7; **37.** $\frac{1}{3}$; **38.** 4; **39.** 37.9

Family Fun

INTEGER FOOTBALL

Players: Two or more

Materials:

- coin
- integer cards
- football field game board
- paper football marker

Directions:

1. Cut out the cards on page FA49, shuffle them, and place them face down.
2. Each team starts on the 50-yard line.
3. A flip of a coin determines who kicks off.
4. The winner of the toss turns over the top card and adds the integer to 50. The result is the number of yards needed to score (negative numbers are desirable!)
5. The ball goes over to the other player or team after each play.
6. A team makes a touchdown and scores six points when the team's total play is zero.
7. To score the point after a touchdown, the team must draw any negative card.
8. After each touchdown, both sides return to the 50-yard line.
9. At half time, usually after 15 minutes, each side must begin at the 50-yard line.
10. The side with the greatest score at the end of the time period, usually 30 minutes, wins.

G 10 20 30 40 50 40 30 20 10 G

G 10 20 30 40 50 40 30 20 10 G

0	1	2	3	4	5	6
7	8	9	10	11	12	13
14	15	16	17	18	19	20
$^{-}1$	$^{-}2$	$^{-}3$	$^{-}4$	$^{-}5$	$^{-}6$	$^{-}7$
$^{-}8$	$^{-}9$	$^{-}10$	$^{-}11$	$^{-}12$	$^{-}13$	$^{-}14$
$^{-}15$	$^{-}16$	$^{-}17$	$^{-}18$	$^{-}19$	$^{-}20$	
$^{-}1$	$^{-}2$	$^{-}3$	$^{-}4$	$^{-}5$	$^{-}6$	$^{-}7$
$^{-}8$	$^{-}9$	$^{-}10$	$^{-}11$	$^{-}12$	$^{-}13$	$^{-}14$
$^{-}15$	$^{-}16$	$^{-}17$	$^{-}18$	$^{-}19$	$^{-}20$	
PENALTY $^{+}5$	PENALTY $^{+}15$	PENALTY $^{-}5$	PENALTY $^{-}15$			

Family Fun Which Letter?

Start at the beginning of each arrow. Use the rule to fill in the boxes. The integer in the box where two arrows cross should satisfy the rule on both arrows.

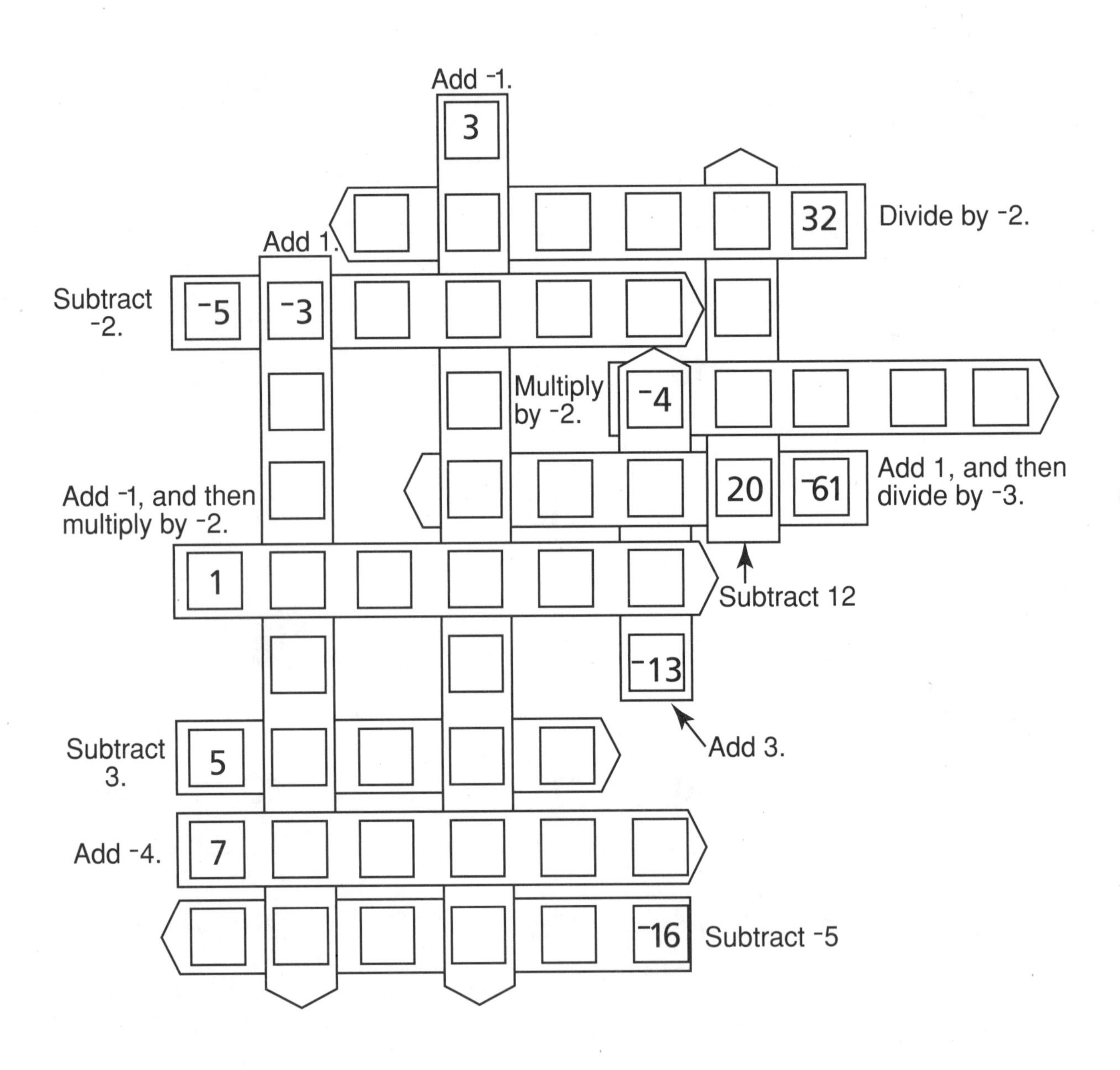

Name

Date

WHAT WE ARE LEARNING

Expressions

Vocabulary

Here are some of the vocabulary words we use in class:

Terms The parts of an algebraic expression separated by an addition or subtraction sign

Square The product of a number and itself

Square root One of the two equal factors of a number

Dear Family,

Your child is learning to write algebraic expressions, to evaluate algebraic expressions, and to evaluate expressions with squares and square roots.

This is how your child is learning to write an algebraic expression for a word expression.

Word Expression	Algebraic Expression
3.5 more than 2 times c	$2c + 3.5$
$\frac{1}{2}$ subtracted from d divided by 4	$\frac{d}{4} - \frac{1}{2}$
the product of h, $5j$, and k	$h \times 5j \times k$
3 less than the quotient of a and 2.1	$\frac{a}{2.1} - 3$

Ask questions such as these as you work together.

How would you write an algebraic expression for these word expressions: 5 more than 11 times x, 7 less than the quotient of y and 12? Your child might respond: the first one is $11x + 5$, and the second one is $\frac{y}{12} - 7$

How would you write word expressions for these algebraic expressions: $\frac{7h}{4} - 10$, $6g + 8.13$? Your child might answer: Ten less than the quotient of 7 times h and 4 for the first one, and 8.13 more than 6 times g for the second one.

Use the steps listed below to help your child evaluate algebraic expressions:

Step 1	Step 2
Combine like terms.	Evaluate the expression.

How would you simplify the following algebraic expressions: $5x + 7x - 20$, $14y - 5y + 11$? Your child might respond: The simplified expressions are $12x - 20$ and $9y + 11$.

Why can it help to simplify an algebraic expression before evaluating it? Your child might say: I will have to perform fewer computations when evaluating the expression if I simplify it first.

Explain each step as you evaluate $5x + 7x - 20$ and $14y - 5y + 11$ for $x = 3$ and $y = 2$. Your child might explain:

$5x + 7x - 20$	Combine the like terms.
$12x - 20$	Replace x with 3.
$12 \cdot \mathbf{3} - 20$	Multiply.
$36 - 20$	Subtract.
16	

$14y - 5y + 11$	Combine the like terms.
$9y + 11$	Replace y with 2.
$9 \cdot \mathbf{2} + 11$	Multiply.
$18 + 11$	Add.
29	

Your child is learning to evaluate expressions containing squares and square roots. These are the steps your child is using.

a.	$\sqrt{16} \times (7 - 3)$	Evaluate $\sqrt{16}$.
	$4 \times (7 - 3)$	Operate inside parentheses.
	4×4	
	16	Multiply.

b.	$2^4 - 10 + \sqrt{9}$	Evaluate 2^4 and $\sqrt{9}$.
	$(16 - 10) + 3$	Operate inside parentheses.
	$6 + 3$	
	9	Add.

Use the practice exercises on the next page to help your child write and evaluate algebraic expressions and expressions with squares and square roots.

As you work with your child, talking about math is the best way to help him or her gain confidence and understanding.

Sincerely,

Name ______________________ PRACTICE/HOMEWORK

Expressions

Simplify.

1. $3x + 5x + 12$ ________
2. $14x + 11x + 22$ ________
3. $63 + 13y - 3y$ ________
4. $99y + 2y - 15$ ________

Write an algebraic expression for the word expression.

5. 5 more than 7 times x ________
6. $\frac{1}{2}$ less than x divided by y ________
7. the quotient of x and 3.5 ________
8. the product of a and $6b$ and c ________
9. the product of 4 and the sum of 6 and y ________
10. the difference of 5 and the product of 9 and n ________

Evaluate the algebraic expression for $x = 2$ and $y = 3$.

11. $15x + 10x - 49$ ________
12. $22x - 11x - 22$ ________
13. $3y - 2y + 13$ ________
14. $8y - 6y + 5$ ________
15. $(x - 2)^2 + 3x$ ________
16. $\sqrt{25} + 2x^2 - 2y$ ________
17. $4y - 2y + 6$ ________
18. $\sqrt{36} + y^2 - 34$ ________
19. $\sqrt{100} - x^2 + \frac{3 \times 2}{\sqrt{49}}$ ________
20. $x^3 - 2y + x(\sqrt{121} + x^2)$ ________
21. $(\sqrt{25} + \sqrt{16})^2 - (\frac{2y^2}{x^2})$ ________
22. $\frac{y^2}{x} + (\sqrt{81} + x^2) - 4y$ ________

Answers: **1.** $8x + 12$; **2.** $25x + 22$; **3.** $63 + 10y$; **4.** $101y - 15$; **5.** $7x + 5$; **6.** $\frac{x}{y} - \frac{1}{2}$; **7.** $\frac{x}{3.5}$; **8.** $a \times 6b \times c$; **9.** $4 \times (6 + y)$; **10.** $5 - 9n$; **11.** 1; **12.** 0; **13.** 16; **14.** 11; **15.** 6; **16.** 7; **17.** 12; **18.** $^-19$; **19.** $6\frac{6}{7}$; **20.** 32; **21.** $76\frac{1}{2}$; **22.** $5\frac{1}{2}$

Family Fun

Pretend that your class is having a pizza party to celebrate the end of the school year. You are the class treasurer and you are responsible for the class funds. Ask a friend or a family member to help you make a diagram to determine the pizzas you will order.

Each person gets three slices of pizza. You have taken orders and found that 11 people want pepperoni only, 8 people want cheese only, 10 people want ham and pineapple only, and 9 people want 1 slice of each.

Pizza	Prices
Medium Round (12 slices)	$10.00
Large Round (16 slices)	$14.00
Extra Large Pan Style (21 slices)	$17.00

1. How many slices of each kind of pizza do you need to order? How many pizzas of each kind should you order?

2. You have $89.00 that your class earned from recycling aluminum cans. Do you have enough money to pay for the pizzas? If not, if each person contributed another $0.50, would you have enough to cover the difference?

3. You get to take home any leftovers. What, if any, would you be taking home?

Answers: 1. 42 slices of pepperoni only; 33 slices of cheese only, 39 slices of ham and pineapple only; 2 extra large pepperoni, 1 extra large cheese, 1 medium cheese, 1 large ham, 2 medium ham; **2.** no; yes; **3.** 1 slice of ham and pineapple

HARCOURT MATH

GRADE 6

Chapter 14

WHAT WE ARE LEARNING

Addition and Subtraction Equations

VOCABULARY

Here are the vocabulary words we use in class:

Equation A statement showing that two quantities are equal

Variable A letter that represents a value you do not know. A variable can be on the left or the right side of an equation.

Subtraction Property of Equality If you subtract the same number from both sides of an equation, the two sides remain equal.

Addition Property of Equality If you add the same number to both sides of an equation, the two sides remain equal.

Identity Property of 0 For all numbers a, $a + 0 = a$.

Commutative Property of Addition The sum stays the same when the order of the addends is changed, $a + b = b + a$.

Name

Date

Dear Family,

Your child is learning to connect words and equations by translating words into numbers, variables, and operations. Once a problem is written as an equation, your child is learning the steps to solve addition and subtraction equations.

Since addition and subtraction are inverse operations, your child will use the inverse operation to get the variable alone on one side of the equation. Here are the steps we use.

Model and write an addition equation.

$n + 6 = 15$

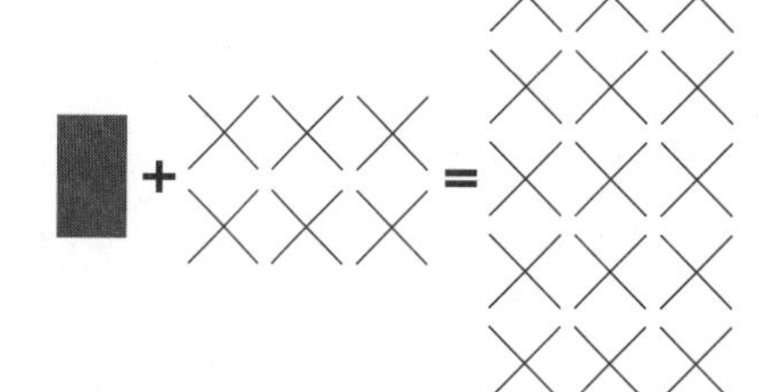

Step 1

To get the variable n alone, you must subtract from each side.

$n + 6 - 6 = 15 - 6$

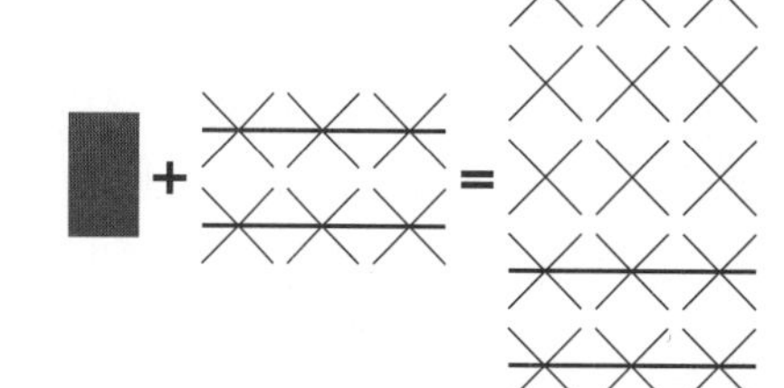

Step 2

What remains is the solution of the equation.

$n + 0 = 9$
$n = 9$

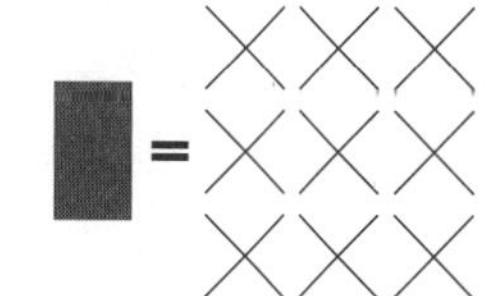

Use the model here and the practice activities on the back of this page to help your child practice solving equations.

Sincerely,

Model and Solve Equations

HOME ACTIVITY

How to model an equation.

1. Draw a green rectangle to represent a variable. The rectangle's shape or size has no relationship to its value; it merely represents an unknown.
2. Draw yellow squares to represent ones in addition problems.
3. Draw red squares to represent ones in subtraction equations.

Solve an addition equation.

1. $n + 5 = 13$

Draw a green rectangle + 5 yellow squares = 13 yellow squares.

Step 1: To get the variable *n* alone, you must subtract the same number from each side. Subtract 5 yellow squares from each side of the equation by crossing out with red marker.

Step 2: What remains is the solution of the equation. What is the value of the green rectangle?

Now try solving these two addition equations:

2. $z + 8 = 19$ **3.** $y + 5 = 16$

Solve a subtraction equation.

4. $n - 3 = 14$

Draw a green rectangle – 3 red squares = 14 yellow squares.

Step 1: To get the variable *n* alone, you must add the same number to each side. Use the yellow marker to draw 3 yellow squares on each side of the equation.

Step 2: What remains is the solution of the equation. What is the value of the green rectangle?

Now try solving these two subtraction equations:

5. $15 = a - 7$ **6.** $x - 6 = 7$

Answers: 1. 8; 2. 11; 3. 11; 4. 17; 5. 22; 6. 13

Name ______________________

PRACTICE/HOMEWORK

Addition and Subtraction Equations

Write an equation for each word sentence.

1. 2.8 more than a number is 4.2. ____________
2. 17 less than a number is 85. ____________
3. The sum of a number and 3.8 is 17.48. ____________
4. $5\frac{1}{4}$ reduced by a number is $3\frac{2}{3}$. ____________
5. Danea grew $2\frac{1}{2}$ inches one summer. This was $1\frac{3}{4}$ inches more than the summer before. How much did she grow the summer before? ____________
6. Carmen spent $16.50 at the amusement park. She had $8.50 left in her wallet. How much money did she start with? ____________

Solve and check.

7. $a + 36 = 98$ ______
8. $87 = b + 29$ ______
9. $16 + c = 53$ ______
10. $12.5 = x + 4.25$ ______
11. $y + 4.3 = 8.1$ ______
12. $6.5 = z + 1.7$ ______
13. $d + 3\frac{1}{5} = 7\frac{9}{10}$ ______
14. $8\frac{4}{5} = e + 5$ ______
15. $9\frac{1}{2} + f = 14\frac{1}{3}$ ______
16. $t - 42 = 117$ ______
17. $u - 28 = 79$ ______
18. $900 = v - 5{,}275$ ______
19. $g - 9.6 = 3.9$ ______
20. $h - 15.6 = 27.8$ ______
21. $34.7 = k - 50.5$ ______
22. $q - \frac{2}{3} = \frac{5}{6}$ ______
23. $4\frac{3}{4} = r - 2$ ______
24. $5.25 + s = 12.6$ ______

Answers: 1. $n + 2.8 = 4.2$; **2.** $n - 17 = 85$; **3.** $n + 3.8 = 17.48$; **4.** $5\frac{1}{4} - n = 3\frac{2}{3}$; **5.** $2\frac{1}{2} - 1\frac{3}{4} = n$; **6.** $m - \$16.50 = \$8.50$; **7.** $a = 62$; **8.** $b = 58$; **9.** $c = 37$; **10.** $x = 8.25$; **11.** $y = 3.8$; **12.** $z = 4.8$; **13.** $d = 4\frac{7}{10}$; **14.** $e = 3\frac{4}{5}$; **15.** $f = 4\frac{5}{6}$; **16.** $t = 159$; **17.** $u = 107$; **18.** $v = 6{,}175$; **19.** $g = 13.5$; **20.** $h = 43.4$; **21.** $k = 85.2$; **22.** $q = 1\frac{1}{2}$; **23.** $r = 6\frac{3}{4}$; **24.** $s = 7.35$

Family Fun

Materials:

- coin
- game cards
- score sheet

Objective: Solve addition and subtraction equations.

Directions:

1. Cut out cards. Shuffle and stack them face down.
2. Player 1 tosses the coin to determine which operation to use. Heads indicates addition, and tails indicates subtraction. The player turns over a card and writes an equation, setting the variable n plus or minus the card value equal to 20.

   Examples: $n + 3.7 = 20$ or $n - 15\frac{1}{4} = 20$.
3. Player 1 solves the equation he or she has created and records the solution as the score.
4. Player 2 takes a turn.
5. Game continues until all the cards are drawn.
6. Players add their scores. The player with the greatest total wins.

Name

Date

Chapter 15

WHAT WE ARE LEARNING

Multiplication and Division Equations

CONCEPTS

Here are some of the concepts we use in class:

Division Property of Equality If you divide both sides of an equation by the same nonzero number, the two sides remain equal.

Multiplication Property of Equality If you multiply both sides of an equation by the same number, the two sides remain equal.

Identity Property of 1
For all numbers a, $1 \times a = a$.

We use this formula:
distance = rate × time, or $d = rt$. If you know two of the three parts of the formula, you can solve for the third part.

Dear Family,

Your child is learning to solve multiplication and division equations, use formulas, and explore two-step equations. Your child is using inverse operations with multiplication and division problems to get the variable alone on one side of the equation.

Here is how we use a formula to solve distance equations.

Find how long it takes to walk one mile, 5,280 feet, at an average speed of 264 feet per minute.

Step 1
Write the formula. distance = rate × time, or $d = rt$

Step 2
Replace d with the distance (5,280) and r with the rate (264). $\mathbf{5{,}280 = 264}t$

Step 3
Use the Division Property of Equality.
$$\frac{5{,}280}{264} = \frac{264t}{264}$$
$$20 = 1t$$

Step 4
Use the Identity Property of 1. $20 = t$

Step 5
Check the solution.
Replace t with 20. $5{,}280 = 264 \times \mathbf{20}$
The solution checks. $5{,}280 = 5{,}280$

Use the model here and the practice activities on the back of this page to help your child practice using a formula to solve equations.

Sincerely,

Use a Formula

1. The three-toed sloth is the world's slowest animal. It moves 6 feet per minute. How long would it take a sloth to walk the length of a football field (100 yards)?

2. A car can travel from San Francisco to Los Angeles in $6\frac{1}{2}$ hours at an average speed of 58 mi per hr. What is the distance between the two cities?

3. A plane can travel from Los Angeles to Honolulu, a distance of 2,557 miles, in $5\frac{1}{2}$ hours. What is its average speed? Round to the nearest whole number.

4. A mountain biker averaged 30 km per hr on a 100 km race. How long did he take? (Round to the nearest tenth of an hour.)

5. A passenger rides the train for $2\frac{1}{2}$ hours and travels 107 miles before disembarking. What is the average speed of the train?

6. A plane leaves St. Louis and flies at an average speed of 850 km per hr for 3 hours. How far does it travel?

Answers: 1. 50 min.; 2. 377 mi; 3. 465 mi per hr; 4. 3.3 hr; 5. 42.8 mi per hr; 6. 2,550 km

Name ______________________________ PRACTICE/HOMEWORK

Multiplication and Division Equations

Write the answer.

1. The ___?___ Property of Equality states that if you multiply both sides of an equation by the same number, the two sides remain equal.
2. The ___?___ Property of Equality states that if you divide both sides of an equation by the same number, the two sides remain equal.

Solve and check. Justify each step.

3. $7x = 49$ _____
4. $9a = 72$ _____
5. $36 = 9r$ _____
6. $\frac{q}{8} = 12$ _____
7. $2.5x = 7.5$ _____
8. $\frac{s}{3} = 135$ _____
9. $256 = 8t$ _____
10. $13.5 = \frac{b}{6}$ _____
11. $^{-}22.8 = 3.8c$ _____

Use the formula $d = rt$ to complete.

12. d = 1,040 km
    r = _____ km per min
    t = 130 min
13. d = 2,500 km
    r = 50 km per sec
    t = _____ sec
14. d = _____ mi
    r = 55 mi per hr
    t = 6 hr

Convert the temperature to degrees Fahrenheit. Write your answer as a decimal.

15. 37°C _____
16. 21.5°C _____
17. 11°C _____
18. 4°C _____

Convert the temperature to degrees Celsius. Write your answer as a decimal and round to the nearest tenth of a degree.

19. 80°F _____
20. 14°F _____
21. 100°F _____
22. 50°F _____

23. The furnace is set to turn on at 62°F. What temperature is that in degrees Celsius? _____
24. A truck maintained a speed of 65 mph for 325 miles. For how many hours did it travel? _____

Answers: 1. Multiplication; **2.** Division; **3.** x = 7; **4.** a = 8; **5.** r = 4; **6.** q = 96; **7.** x = 3; **8.** s = 405; **9.** t = 32; **10.** b = 81; **11.** c = -6; **12.** 8; **13.** 50; **14.** 330; **15.** 98.6 °F; **16.** 70.7 °F; **17.** 51.8 °F; **18.** 39.2 °F; **19.** 26.7 °C; **20.** -10 °C; **21.** 37.8 °C; **22.** 10 °C; **23.** 16.7 °C; **24.** t = 5 hr

MATH GAME

Family Fun

Sea-Mail

Solve each equation. Then put the letter of the variable above its value to answer the riddle.

$14d = 84$; $d =$ ______________

$88 = 8y$; $y =$ ______________

$\frac{o}{4} = 6$; $o =$ ______________

$\frac{r}{2} = 9$; $r =$ ______________

$15c = 75$; $c =$ ______________

$12s = 108$; $s =$ ______________

$\frac{n}{0.5} = 8$; $n =$ ______________

$\frac{b}{3} = 4$; $b =$ ______________

$180 = 12e$; $e =$ ______________

How did the Vikings send secret messages?

____ ____   ____ ____ ____ ____ ____   ____ ____ ____ ____

12 11   4 24 18 9 15   5 24 6 15

Answer: by Norse code

Name

Grade 6

Chapter 16

WHAT WE ARE LEARNING

Geometric Figures

VOCABULARY

Here are the vocabulary words we use in class:

Ray Part of a line that has one endpoint

Vertex The point where two or more rays meet

Angles:

Vertical two angles that are congruent and opposite each other when two lines intersect

Adjacent Side-by-side angles with a common vertex and ray

Complementary Two angles whose measures have a sum of 90°

Supplementary Two angles whose measures have a sum of 180°

Lines:

Parallel Lines in a plane that are always the same distance apart

Perpendicular Lines that intersect to form 90° angles, or right angles

Date

Dear Family,

Your child is learning to solve problems involving figures in two and three dimensions; to classify parallel, intersecting, and perpendicular lines; to name, measure, and draw angles; and to explore relationships among angles.

Here are the relationships we use to find the unknown measures of certain angles:

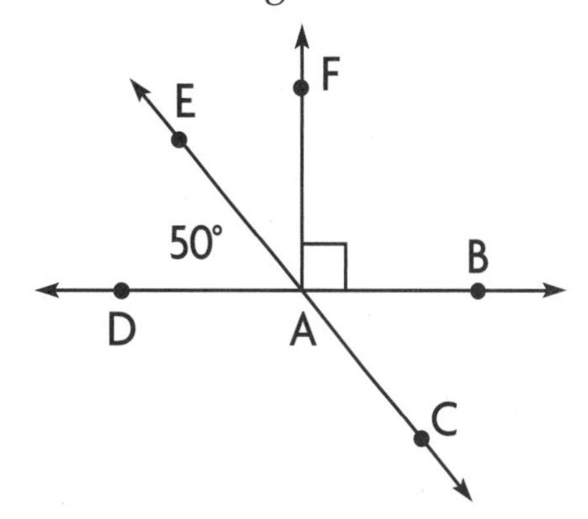

Vertical angles are formed opposite each other when two lines intersect; they have the same measure, so they are always congruent. Angle *BAC* and angle *EAD* are vertical angles.

Can you tell the measure of angle *BAC*? Your child might say: Angle *BAC* has the same measure as angle *EAD*; therefore, it is 50°.

Adjacent angles are side-by-side and have a common vertex. Angle *EAF* and angle *EAD* are adjacent angles.

If angles are complementary, their measures have the sum of 90°. Angle *EAF* and angle *EAD* are complementary angles.

If two adjacent angles are supplementary, their measures have the sum of 180°. Angle *BAC* and angle *CAD* are supplementary angles.

Use the model here and the practice activities on the back of this page to help your child practice recognizing angle relationships and finding unknown measures.

Sincerely,

Angle Solver

HOME ACTIVITY

Use what you know about the relationships among the measures of angles to find the unknown angle measure.

1. 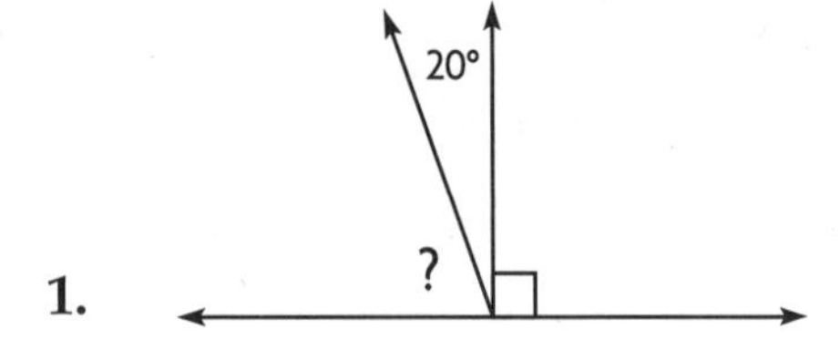

2. 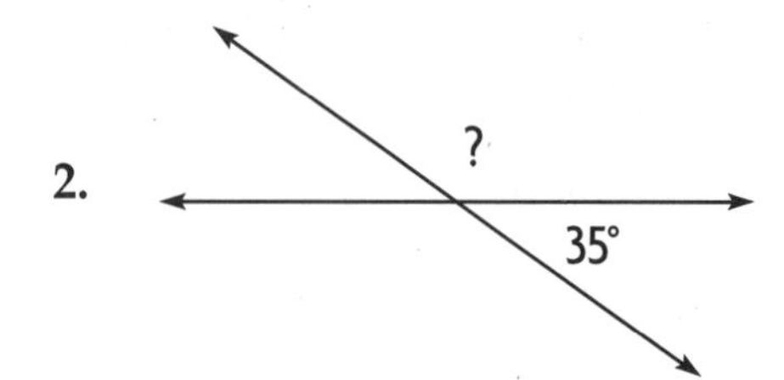

3. 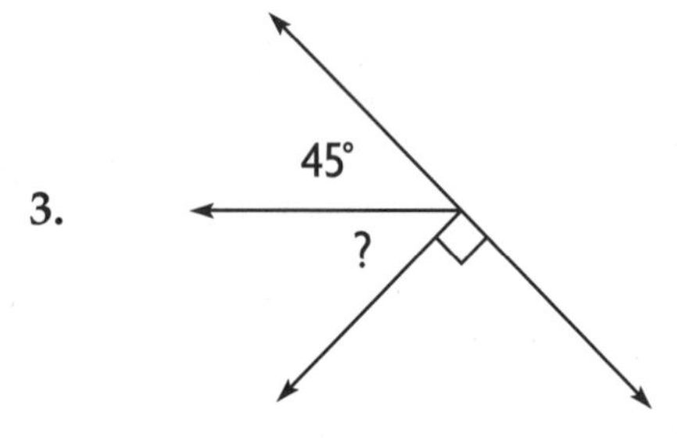

4. 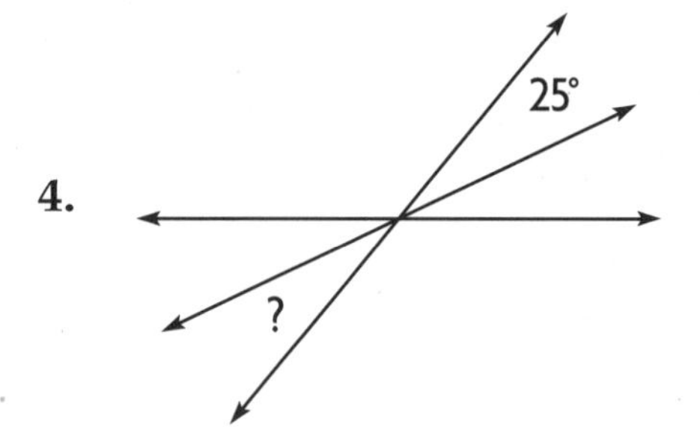

5. 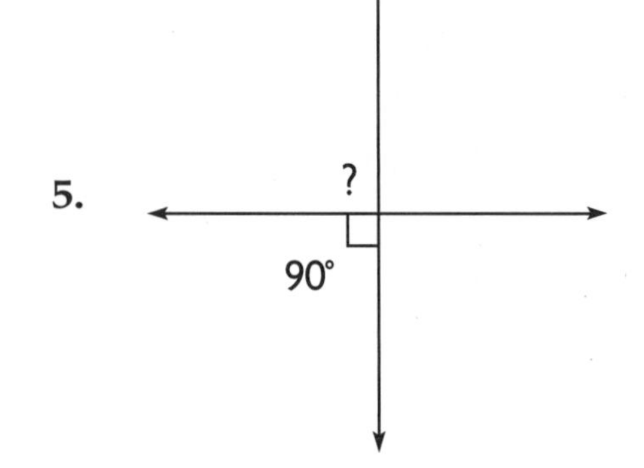

6. 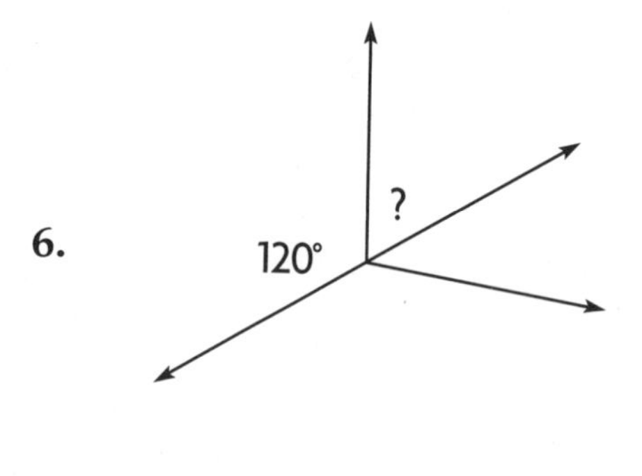

Answers: 1. 70°; 2. 145°; 3. 45°; 4. 25°; 5. 90°; 6. 60°

Name ______________________________ **PRACTICE/HOMEWORK**

Geometric Figures

Name the geometric figure.

• S

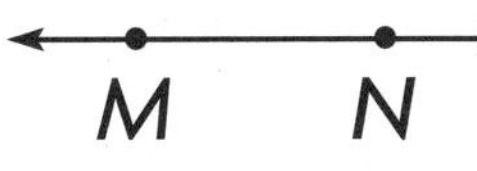

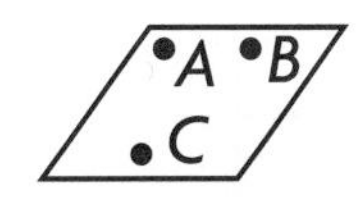

1. ________ 2. ________ 3. ________

R Q

4. ________ 5. ________

Tell if the angle is acute, right, obtuse, or straight.

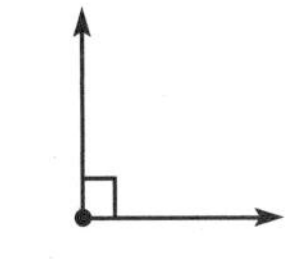

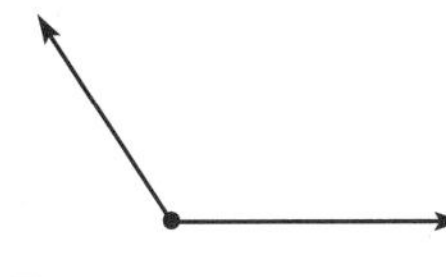

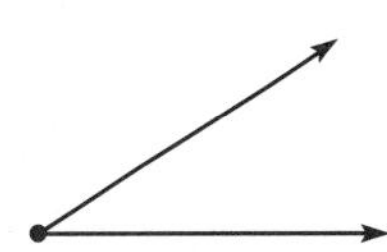

6. ________ 7. ________ 8. ________

Tell if the angles are vertical, adjacent, complementary, supplementary, or have no relationship.

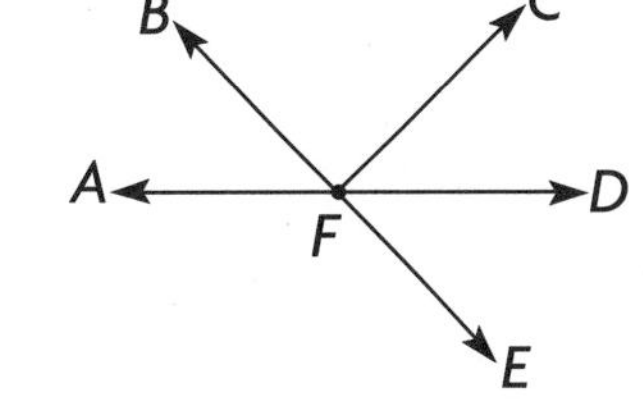

9. Angle *AFC* and angle *CFD* __________
10. Angle *AFB* and *EFD* __________
11. Angle *BFC* and angle *EFD* __________

Use the figure to name the lines.

12. Name a line that is parallel to line *ED*. ____
13. Name all the lines that are perpendicular to and intersect line *AB*. ____
14. Name a line that is parallel to line *AD*. ____
15. Name all the lines that intersect line *EC*.

____

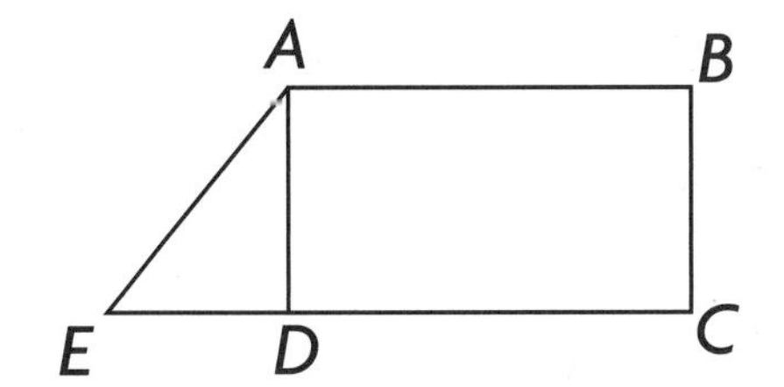

Answers: 1. point *S*; **2.** line *MN*; **3.** plane *ABC*; **4.** line segment *QR* or *RQ*; **5.** ray *EF*; **6.** right; **7.** obtuse; **8.** acute; **9.** adjacent and supplementary; **10.** vertical; **11.** no relationship; **12.** line *AB*; **13.** line *AD* and line *BC*; **14.** line *BC*; **15.** line *AD*, line *AE*, and line *BC*

Family Fun Giving ☞ Directions

MATH GAME

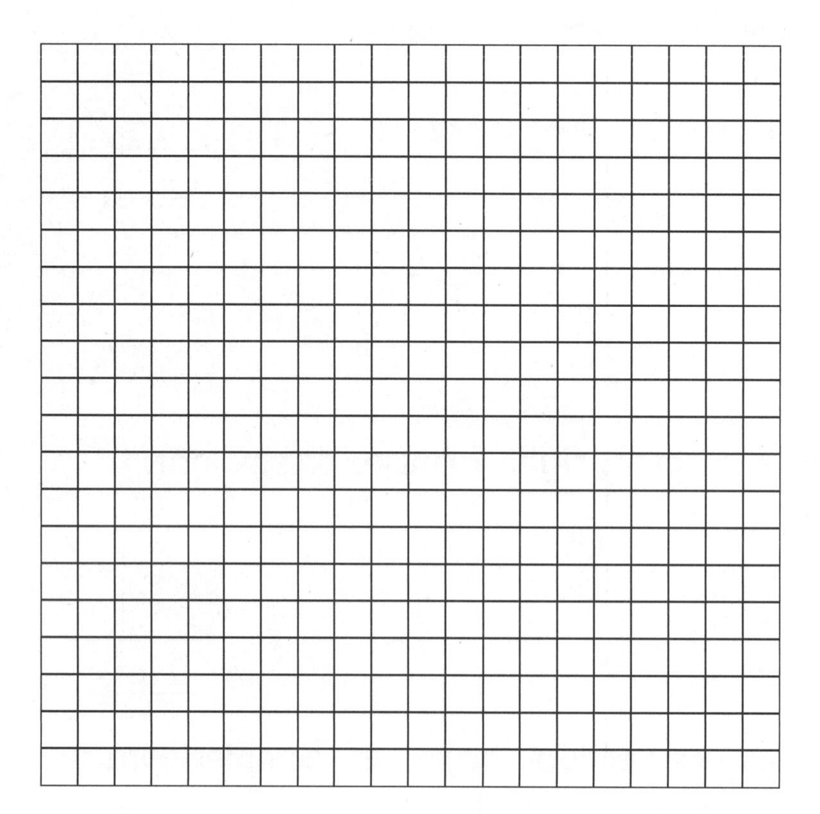

Use the grid to design an amusement park with 6 to 8 fantastic rides. Then name the streets and rides on the grid. Write directions to each ride from the front gate using the words parallel, perpendicular, and intersecting.

Example

To The Wave: Follow First Street until it intersects Beach Street, which is perpendicular to it. Go up Beach Street, which is perpendicular to First Street, until it intersects with Third Street. Turn left on Third Street, and The Wave will be on your right. Happy surfing!

Name

GRADE 6

Date

Chapter 17

WHAT WE ARE LEARNING

Plane Figures

VOCABULARY

Here are the vocabulary words we use in class:

Acute triangle A triangle that contains only acute angles

Obtuse triangle A triangle that contains one obtuse angle

Right triangle A triangle that contains one right angle

Quadrilateral A polygon with four sides and four angles. These are examples of quadrilaterals: parallelogram, rectangle, rhombus, square, trapezoid

Radius Any line segment with one endpoint at the center of a circle and the other endpoint on the circle

Diameter Any line segment that passes through the center of a circle and has both endpoints on the circle

Chord A line segment with endpoints on a circle

Arc Part of a circle, identified by its endpoints

Dear Family,

Your child is learning to find measures of angles in triangles and to solve problems involving quadrilaterals. Your child is interpreting the vocabulary related to plane figures to draw two-dimensional figures and to identify parts of circles.

Here is an activity you can use with your child to classify five types of quadrilaterals.

Look at the models on the next page and complete the chart as you follow the steps below to classify each figure.

- **STEP 1**
  Select one of the five quadrilaterals. Find the row for that figure in the chart. In the second column, describe characteristics of the lengths of the sides.
- **STEP 2**
  In the third column, describe characteristics of the measures of the angles.
- **STEP 3**
  In the fourth column, describe any parallel sides.
- **STEP 4**
  Under the chart write a descriptive statement about the quadrilateral, using the format "If the quadrilateral has ______, then it is a ______."

Encourage your child to do this activity with you and other members of your family.

Sincerely,

Describing Quadrilaterals

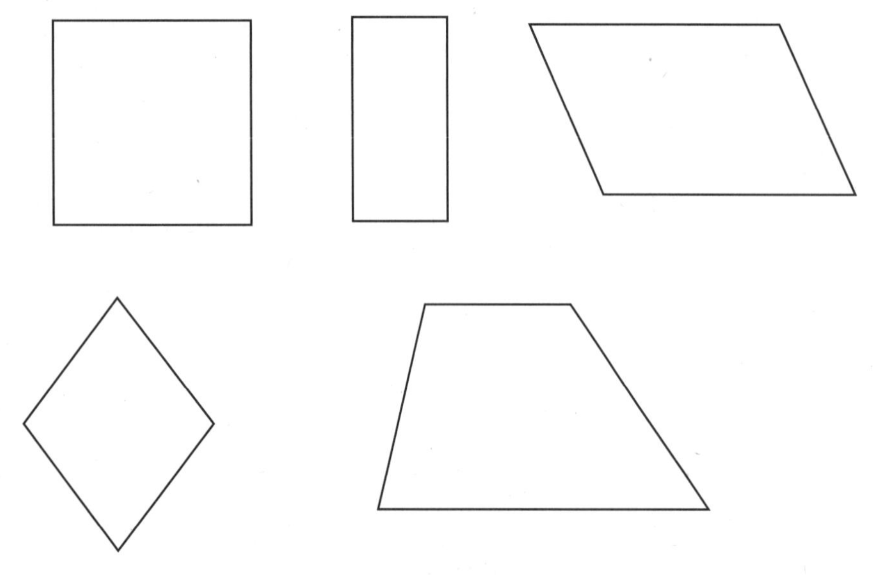

Use the models of the quadrilaterals to help you fill out the chart below. Then describe each quadrilateral, using the "if...then" format.

Quadrilaterals			
Type	**Length of Sides**	**Angle Measures**	**Parallel Sides**
Rectangle			
Square			
Rhombus			
Parallelogram			
Trapezoid			

Descriptions of Quadrilaterals

Rectangle ______________________________

Square ______________________________

Rhombus ______________________________

Parallelogram ______________________________

Trapezoid ______________________________

Answers: Check students' work.

Name ______________________ PRACTICE/HOMEWORK

Plane Figures

Find the measure of the angle and classify the triangle.

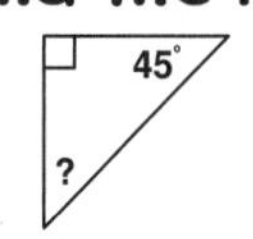

1. ________

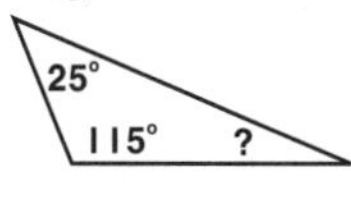

2. ________

Give the most exact name for the figure.

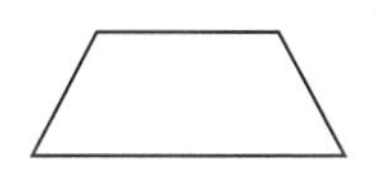

3. ______________

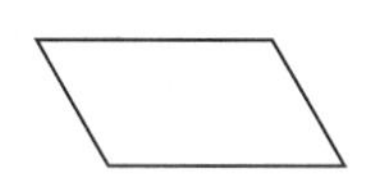

4. ______________

5. ______________

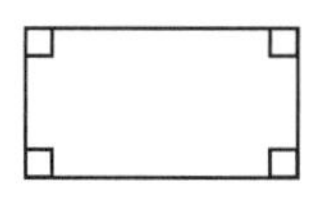

6. ______________

Complete the statement, giving the most exact name for the figure.

7. If a parallelogram has four right angles and four congruent sides, then it is a __________.

8. If a quadrilateral has exactly one pair of parallel sides, then it is a __________.

9. If a quadrilateral has two pairs of parallel and congruent sides, and no right angles, then it is a __________.

10. If a parallelogram has four congruent sides and no right angles, then it is a __________.

Draw the figure.

11. a parallelogram with four right angles and opposite sides congruent

12. an acute triangle with no congruent sides

13. a circle with a chord

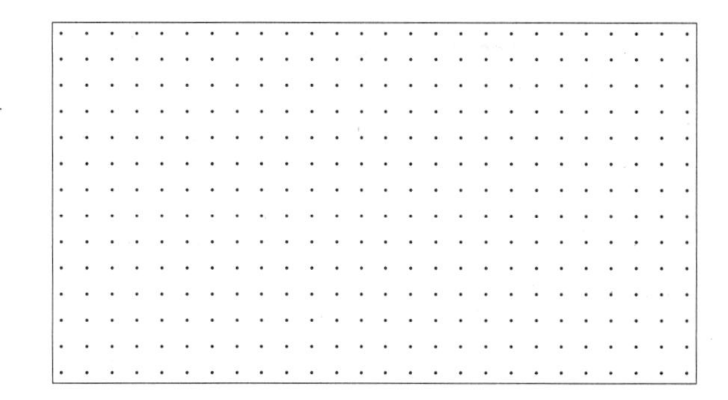

Name the given parts of the circle.

14. center ______

15. radii ______

16. diameter ______

17. chord ______

18. arc ______

19. sector ______

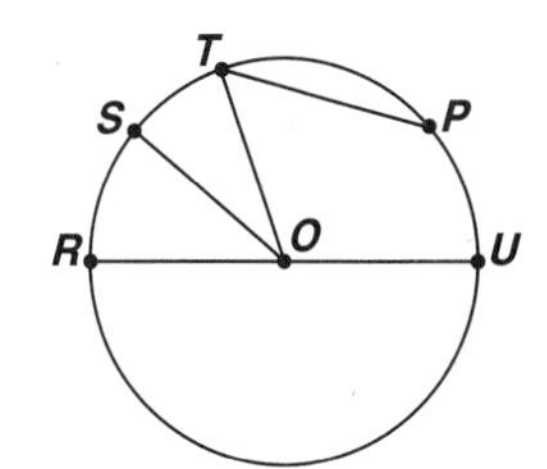

Answers: 1. 45°; right; 2. 40°; obtuse; 3. trapezoid; 4. parallelogram; 5. rhombus; 6. rectangle; 7. square; 8. trapezoid; 9. parallelogram; 10. rhombus; For 11–13, check students' drawings; 14. O; 15. $\overline{OU}$, $\overline{OT}$, $\overline{OS}$, $\overline{OR}$; 16. $\overline{RU}$; 17. $\overline{TP}$, or $\overline{RU}$; 18. Possible answer: $\overset{\frown}{ST}$; 19. Possible answer: sector ST

Materials

- number cube
- paper and pencil

Objective Be the first to equip your "universe" with a "sector." The sector of a circle is the region enclosed by two radii and an arc.

Directions

Each person draws a "universe" (a large circle).

Players take turns rolling the number cube. The chart shows what you draw. Label all points.

Number rolled	Draw
1	center
2	diameter
3	radius
4	point
5	arc (two points)
6	chord

Rules

Players cannot draw a circle part unless the prerequisite is there. For example, players may not draw a diameter or radius unless the center is there, or a chord unless an arc or two points are there.

Since the game continues until one player makes a sector, there may be multiples of some circle parts, such as diameters or points. Players are encouraged to use strategy to decide where to place their drawings of circle parts. Remember, since the objective is to be the first to make a sector, work towards having two radii and an arc (named by its two endpoints).

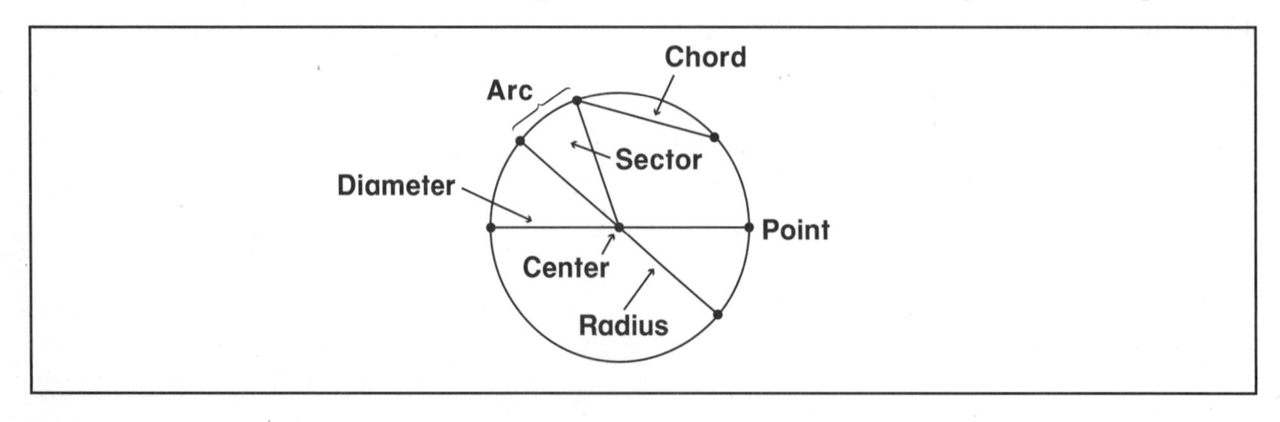

HARCOURT MATH

GRADE 6

Chapter 18

WHAT WE ARE LEARNING

Solid Figures

VOCABULARY

Here are the vocabulary words we use in class:

Polyhedron A solid figure with flat faces that are polygons

Lateral faces The flat or curved surfaces of a solid figure that form its sides

Bases The plane figures that form the top and bottom of a cylinder or prism, or the bottom of a pyramid or cone

Vertex The point of intersection of three or more edges of a solid figure; the top (or bottom point) of a cone.

Prism A solid figure named for the shape of its bases

Net A pattern of plane figures that can be folded to form a specific solid figure

Name

Date

Dear Family,

Your child is learning about solid figures. Your child will use correct vocabulary to identify and describe solid figures, then draw the figures from different views.

Here is how we identify a solid figure by using different views.

Name the solid figure that has the given views.

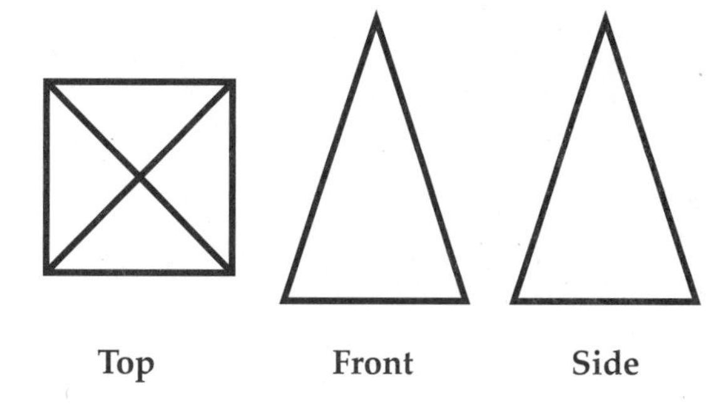

- **STEP 1**
  Identify the base by looking at the top view or top and bottom view.
- **STEP 2**
  Identify the lateral faces by looking at the front and side views.
- **STEP 3**
  Remember the solid figure whose bases and lateral faces match what you have identified.

Now, draw a variation of this figure. Draw the top, front, and side views.

Use the model here and the practice activities on the back of this page to help your child practice identifying solid figures by using different views. Encourage your child to do this activity with you and other members of your family.

Sincerely,

The View is Everything

HOME ACTIVITY

Name and sketch the solid figure shown.

1.

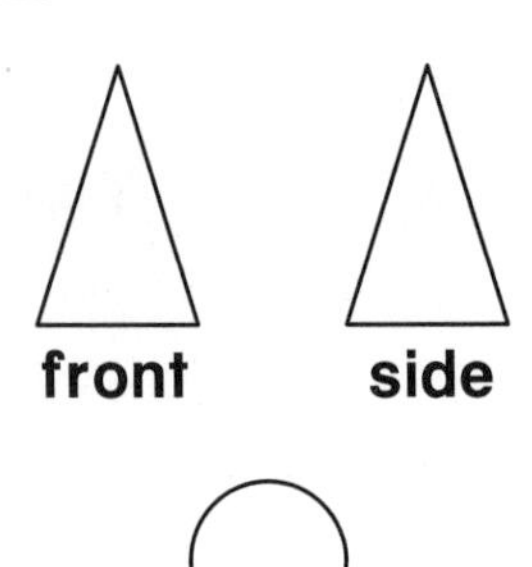

2.

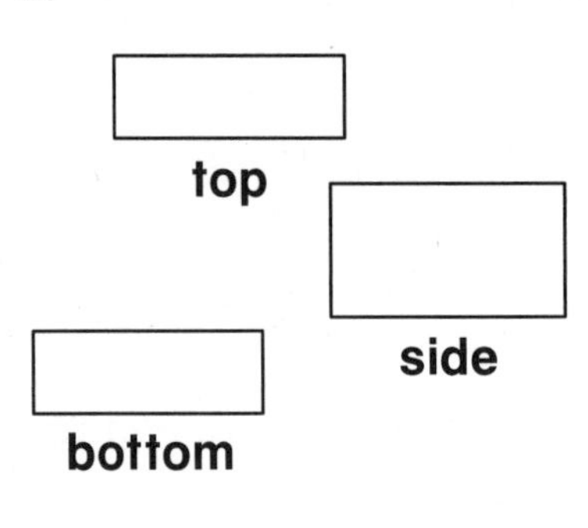

3.

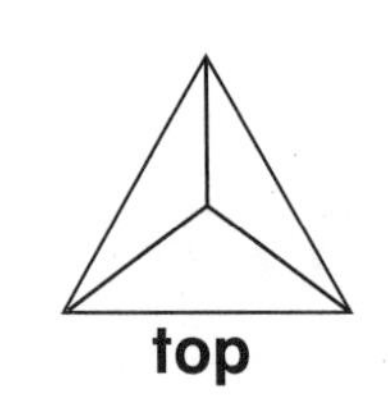

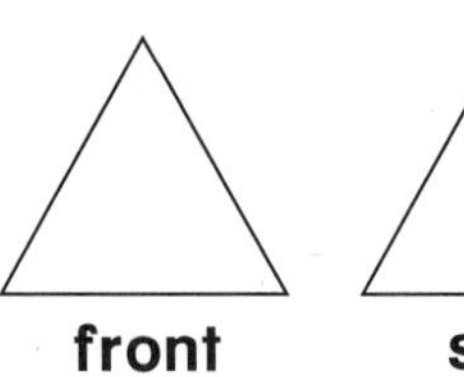

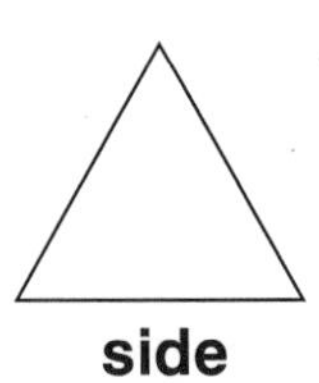

This solid figure is made by combining cubes. Draw the top, front, and side views.

4.

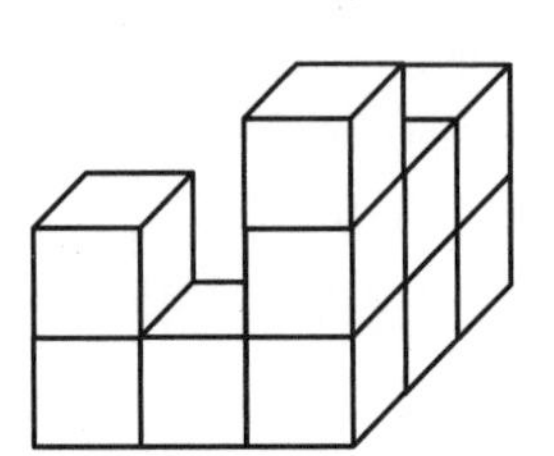

5.

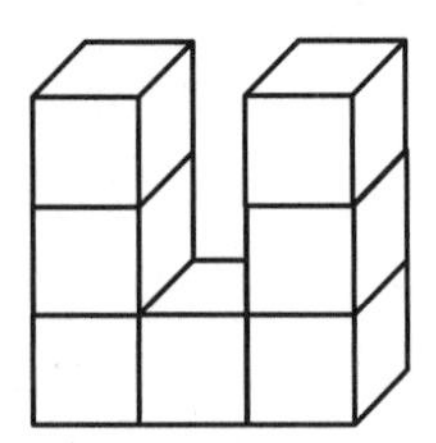

Answers: 1. cylinder; check students' sketches; 2. rectangular prism; check students' sketches; 3. triangular pyramid; check students' sketches 4. top side front ; 5. top side front

Name ____________________

Solid Figures

Classify the figure. Is it a polyhedron?

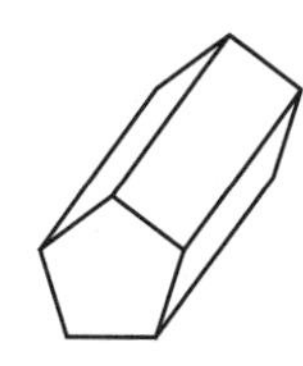

1. ________

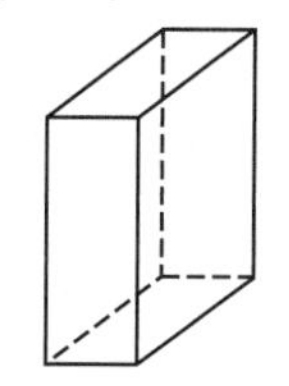

2. ________

3. ________

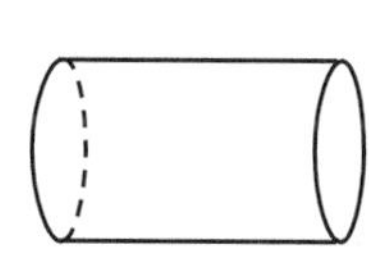

4. ________

Name the solid figure that has the given views.

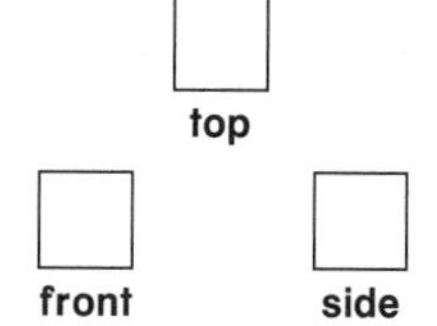

5. ________

top
bottom
side

6. ________

top
bottom
side

7. ________

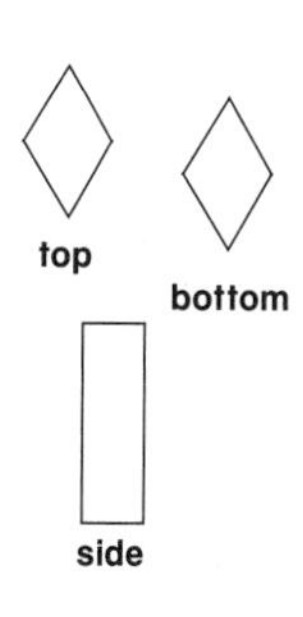

8. ________

Draw the front, top, and side view of each solid.

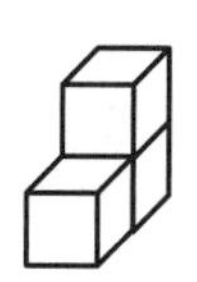

9. ________

10. ________

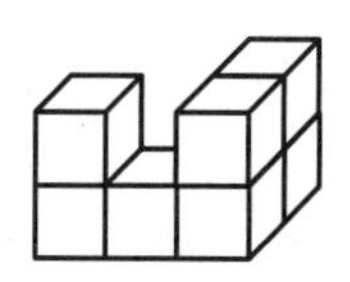

11. ________

12. ________

Answers: 1. pentagonal prism; yes; 2. rect. prism; yes; 3. cone; no; 4. cylinder; no; 5. square prism; 6. cylinder; 7. triangular prism; 8. rectangular prism; 9. top side front; 10. top side front; 11. top side front; 12. top side front

Family Fun

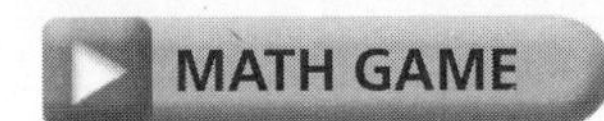

Net Knowledge

Identify the following solid figures by their nets.

3 cubes, 2 rectangular prisms, 1 triangular prism, 3 pyramids, 1 cone, 1 cylinder.

1.

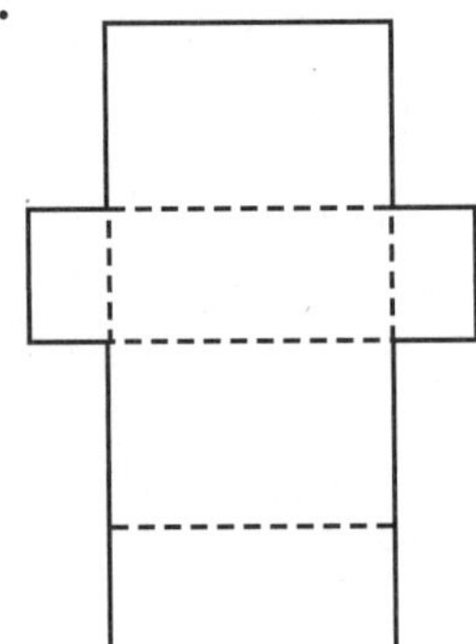

2.

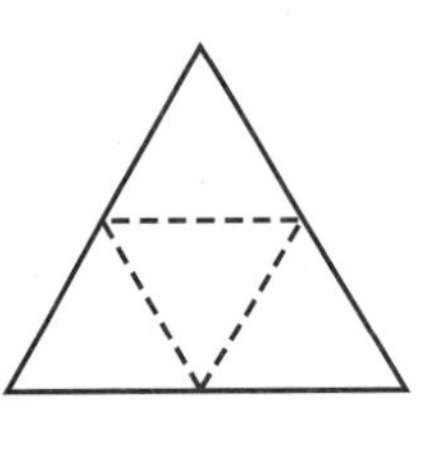

3.

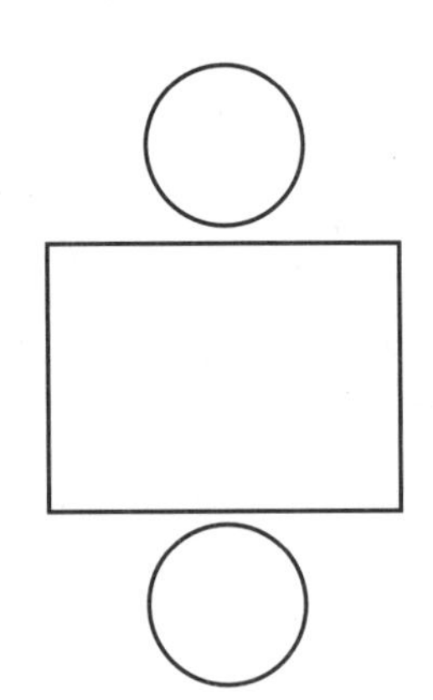

4.

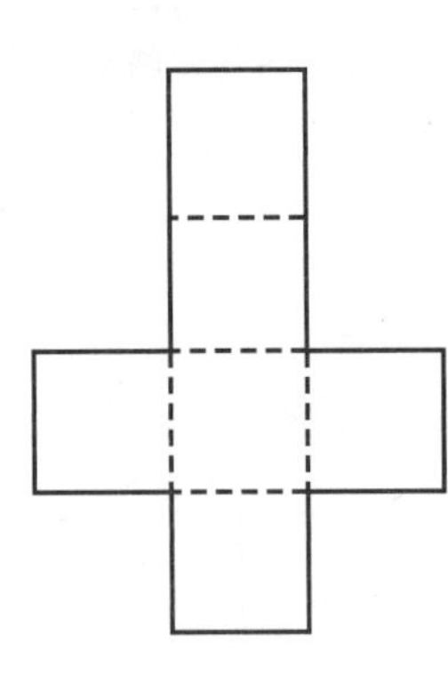

5.

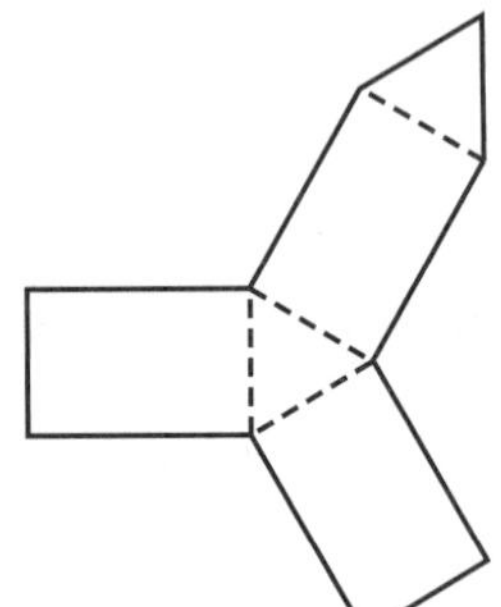

6.

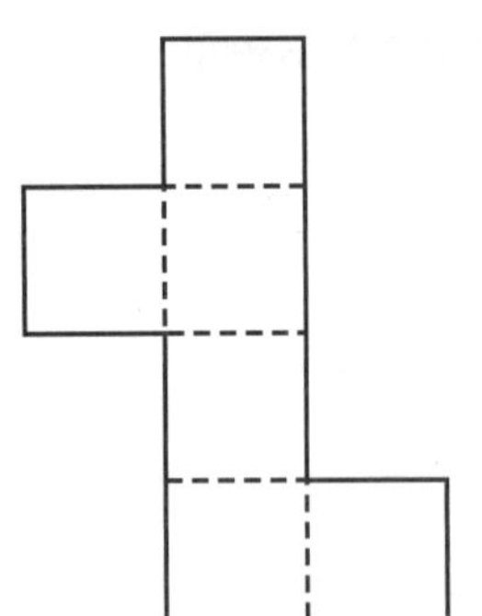

7.

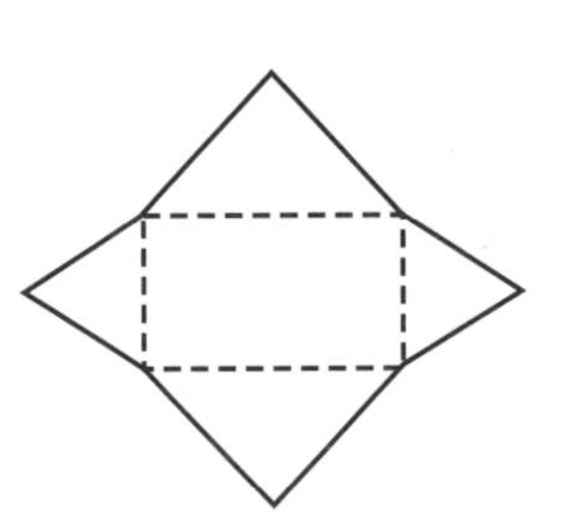

9.

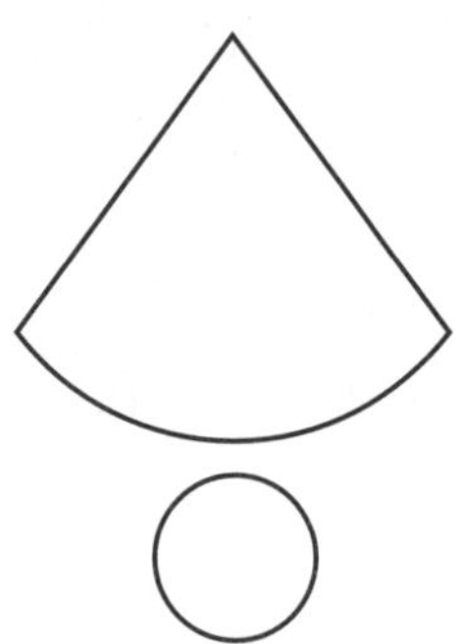

9.

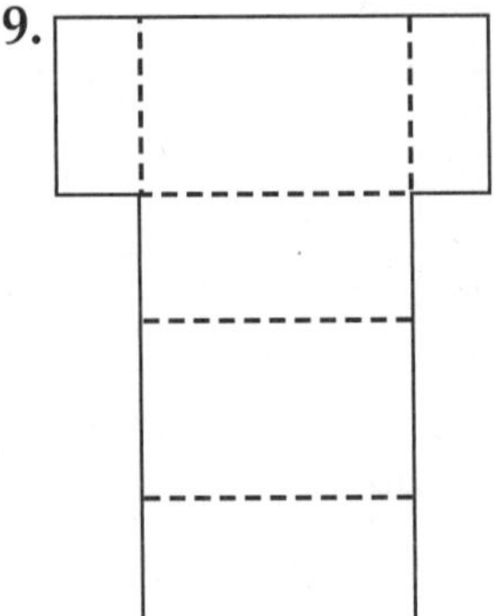

10.

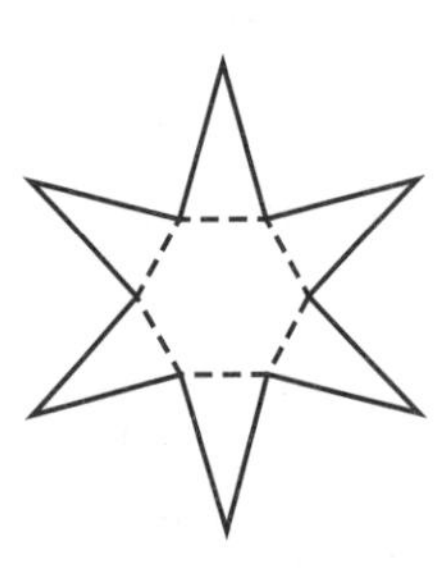

11.

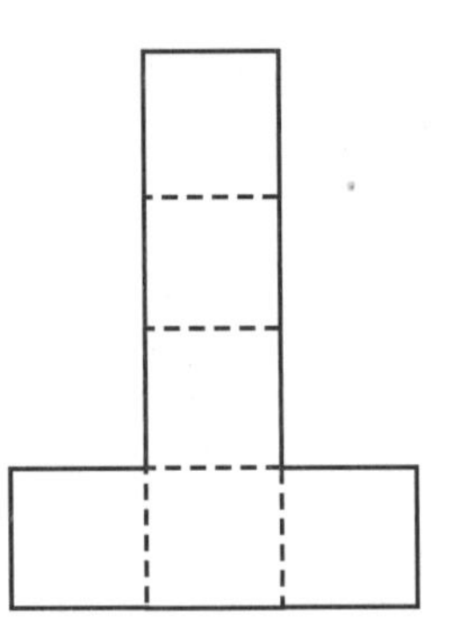

Answers: 1. rectangular prism; 2. pyramid; 3. cylinder; 4. cube; 5. triangular prism; 6. cube; 7. pyramid; 8. cone; 9. rectangular prism; 10. pyramid; 11. cube

HARCOURT MATH

GRADE 6

Chapter 19

WHAT WE ARE LEARNING

Congruence and Similarity

VOCABULARY

Here are some of the vocabulary words we use in class:

Bisect To divide into two congruent parts

Midpoint The point halfway between the endpoints of a line seqment

Similar Figures Figures with the same shape but not necessarily the same size

Congruent Having the same size and shape

Name

Date

Dear Family,

Your child is beginning to study similar and congruent figures. It is important that your child knows the difference between similar figures and congruent figures.

Your child learned that similar figures have the same shape while congruent figures have the same shape and the same size. Examples of similar figures and congruent figures are shown below.

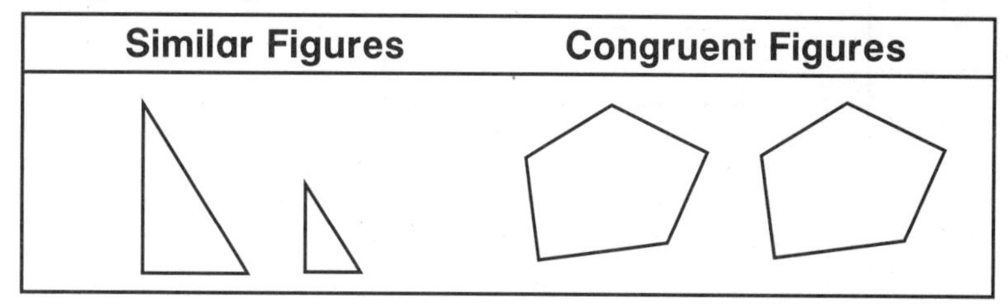

Figures that are congruent are also similar. However, similar figures are not necessarily congruent. Figures can be congruent, similar, both, or neither.

Tell if the figures in each pair appear to be similar, congruent, both, or neither.

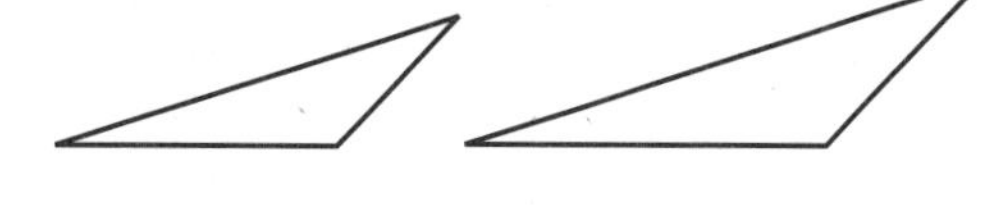

The figures have the same shape so they appear to be similar.

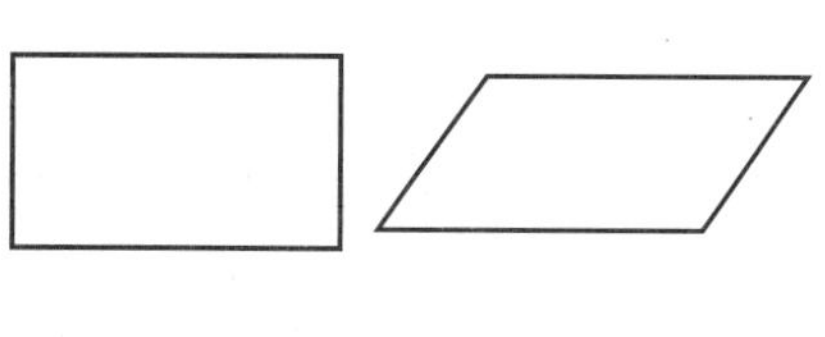

The figures do not have the same shape or size so they are neither similar nor congruent.

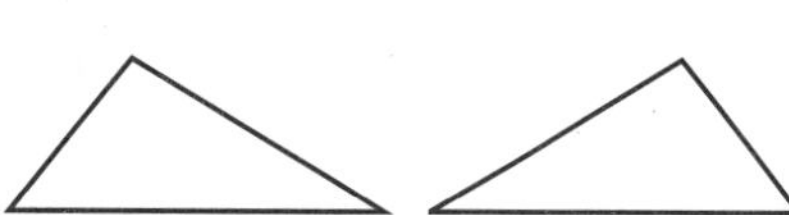

The figures are the same shape and the same size so they are congruent. Since all congruent figures are similar, these figures are both congruent and similar.

Use the models here and the practice activities on the back of this page to help your child practice identifying similar and congruent figures. Encourage your child to do this activity with you and other members of your family.

Sincerely,

Build Your Own Rectangles

Cut out the rectangles below. Answer each question using the rectangles.

1. How many different pairs of congruent rectangles can you build?
2. How many different pairs of similar rectangles can you build?

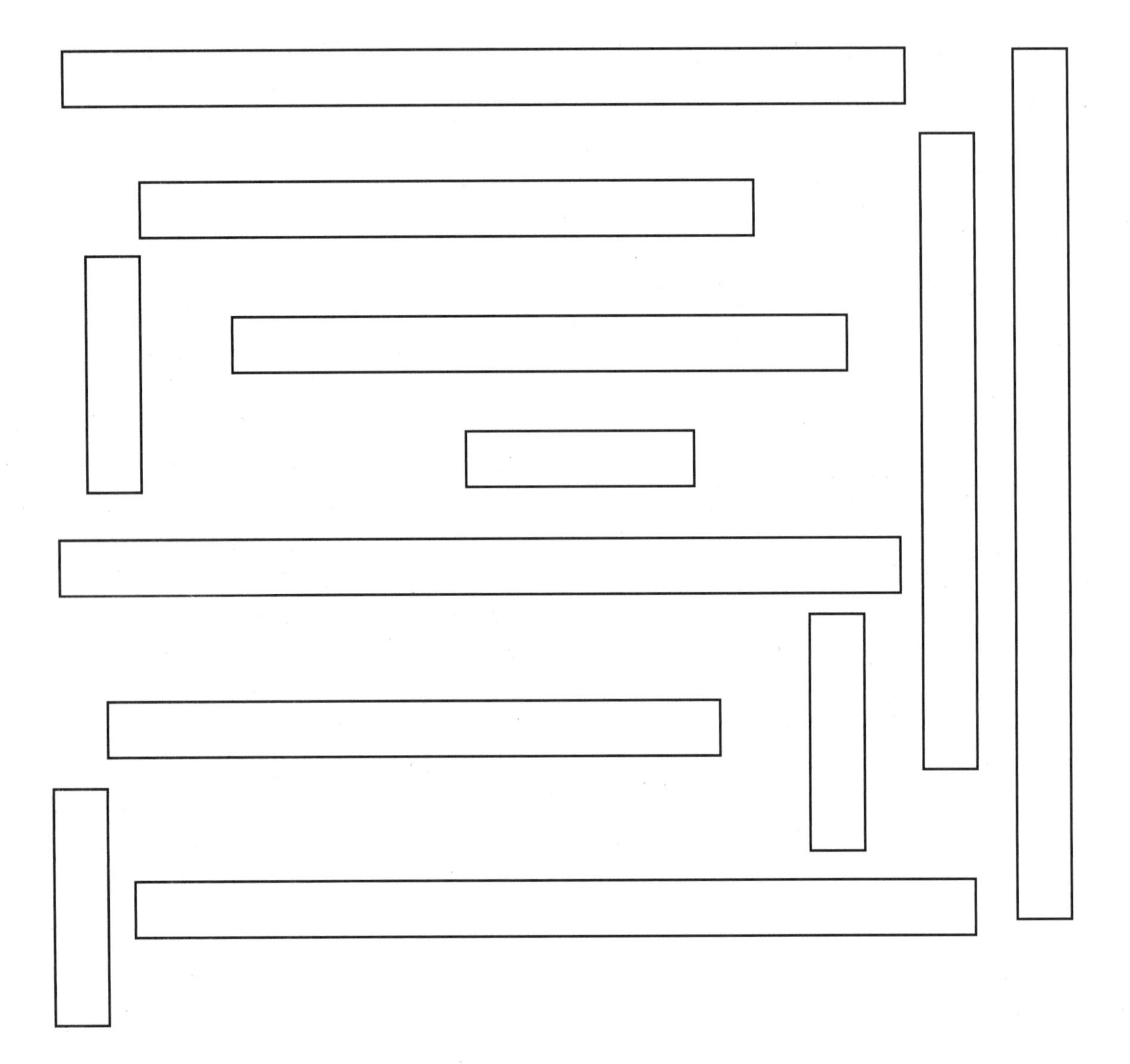

Answers: 1. Check students rectangles. 2. Check students rectangles.

Name ______________________________ **PRACTICE/HOMEWORK**

Find the measure of each angle using a protractor. Then tell whether the angles in each pair are congruent. Write *yes* or *no*.

1. 

2. 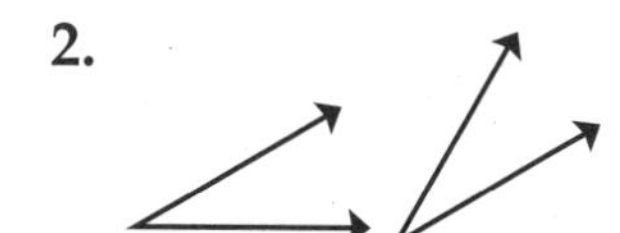

3. 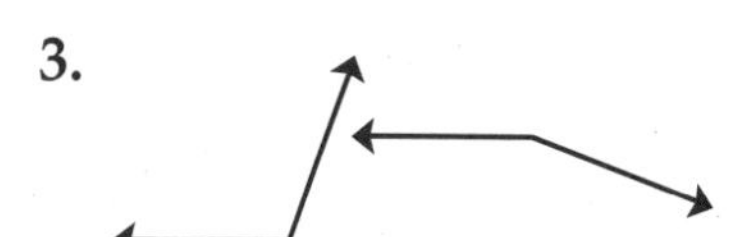

Use a compass to determine which line segments are congruent. Circle all congruent line segments.

4. 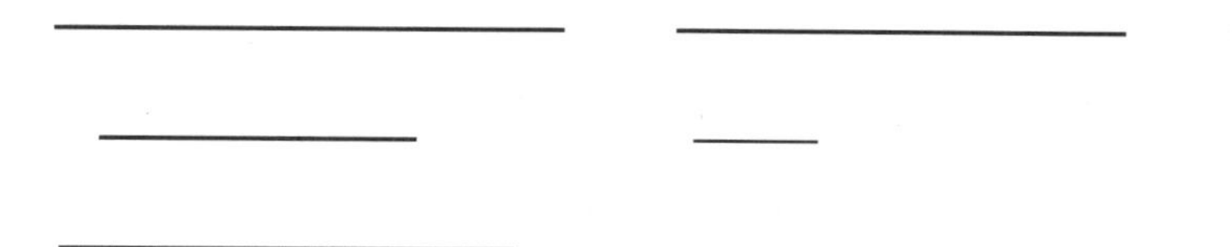

Use a compass and a straightedge to construct an angle congruent to the given angle. Then bisect the angle.

5. 

6. 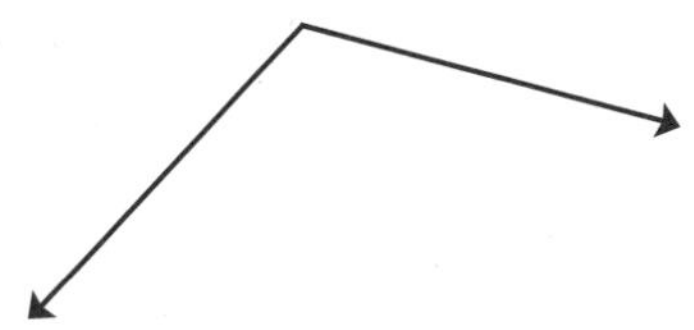

7. 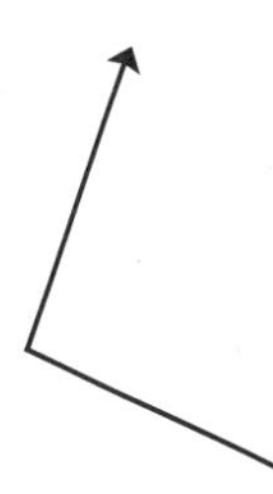

Tell whether the figures in each pair appear to be similar, congruent, both, or neither.

8. 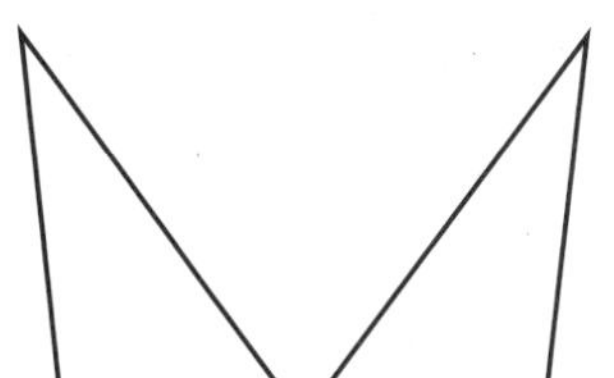

9. 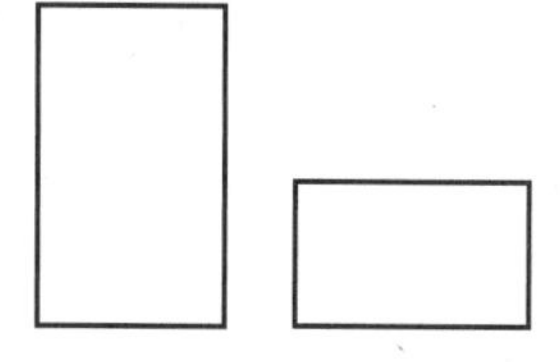

10. 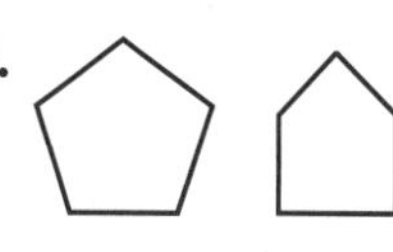

Construct a line parallel to the given line.

11. 

12. 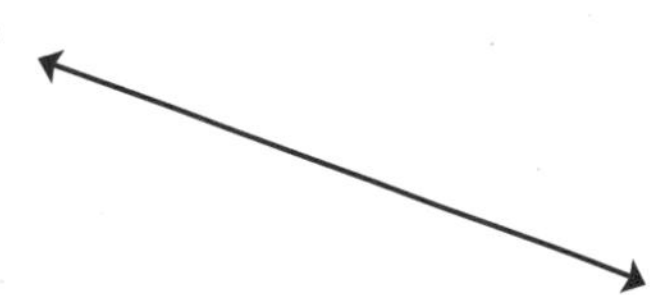

Answers: 1. 90°, yes; 2. 30°, yes; 3. 110°, 160°, no; 4. four of the line segments are congruent; 5–7. Check students' constructions.; 8. both; 9. similar; 10. neither; 11–12. Check students' constructions.

Player 1 chooses one angle from those shown below. Then player 1 constructs a congruent angle on the line at the bottom of the page and extends the top ray of the angle to see how many meteors will be hit by the ray. Player 1 earns the total value of each of the meters that are hit. Once a meteor has been hit it cannot be hit again. Players take turns until all meteors have been hit. The player with the most points is the winner.

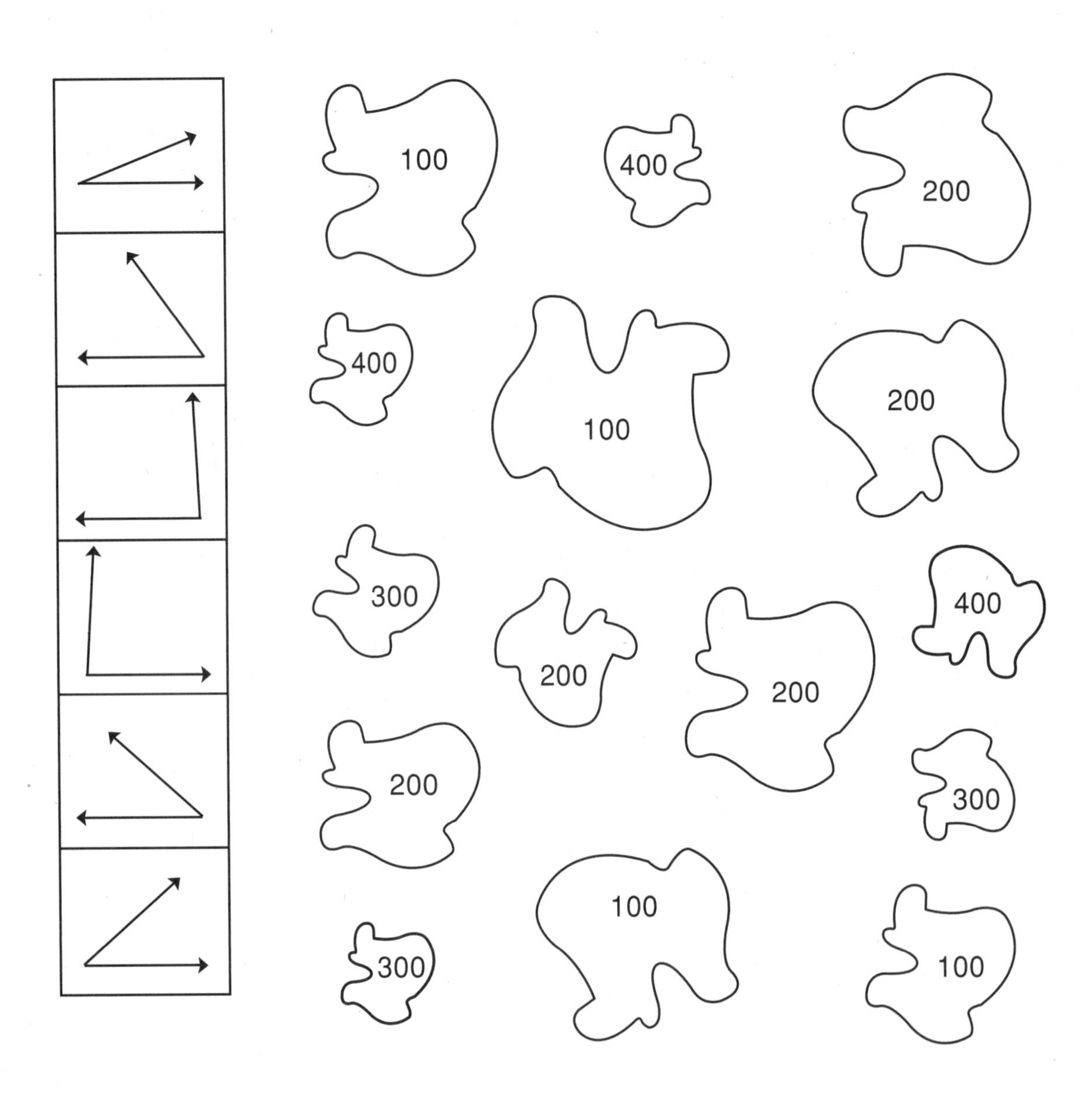

HARCOURT MATH

GRADE 6

Chapter 20

WHAT WE ARE LEARNING

Ratio and Proportion

VOCABULARY

Here are the vocabulary words we use in class:

Ratio A comparison of two numbers, a and b, that can be written as a fraction $\frac{a}{b}$

Proportion An equation that shows two equivalent ratios

Rate A ratio that compares two quantities having different units of measure

Similar figures Figures that have the same shape, but may or may not have the same size

Corresponding sides In similar figures, corresponding sides have the same ratio.

Corresponding angles In similar figures, corresponding angles are congruent.

Scale A ratio between two sets of measurements

Name

Date

Dear Family,

Your child is learning about ratios and proportions. We use ratios and proportions to solve problems. We use ratios to identify similar figures and proportions to find unknown measures.

Here is how we use proportions and indirect measurement to find the height of objects that are difficult to measure directly. Find the height of a tree by using similar figures.

- **STEP 1**

  Write a proportion.

  Write the ratio of the corresponding shadows:

  $\frac{\text{my shadow}}{\text{tree's shadow}} = \frac{2 \text{ ft}}{10 \text{ ft}}$

  Write the ratio of the corresponding heights:

  $\frac{\text{my height}}{\text{tree's height}} = \frac{5 \text{ ft}}{h \text{ ft}}$

- **STEP 2**

  Find the cross products.

  $\frac{2}{10} = \frac{5}{h}$

  $2 \times h = 5 \times 10$

- **STEP 3**

  Solve the equation.

  $2h = 50$

  $\frac{2h}{2} = \frac{50}{2}$

  $h - 25$

So, the tree is 25 feet tall.

Use the model here and the practice activities on the back of this page to help your child practice indirect measurement. Encourage your child to do this activity with you and other members of your family.

Sincerely,

Indirect Measurement

Use indirect measurement to find the height of objects in your neighborhood. On a sunny day, stand next to an object of unknown height. Measure the length of your shadow and that of the other object. Then use your data to find the unknown height.

Use the data to find the unknown height.

1. A yardstick 3 ft long with a shadow of 7 ft
   A wall of unknown height with a shadow of 28 ft

2. A student 1.5 m tall with a shadow of 2.5 m
   A building of unknown height with a shadow of 37.5 m

3. A student 5 ft tall with a shadow of 3 ft
   A street lamp of unknown height with a shadow of 18 ft

4. A student 5 ft tall with a shadow of $1\frac{1}{2}$ ft
   A water tower of unknown height with a shadow of 12 ft

5. A signpost 4 m tall with a shadow of 2 m
   A flagpole of unknown height with a shadow of 8 m

Answers: 1. 12 ft; 2. 22.5 m; 3. 30 ft; 4. 40 ft; 5. 16 m

Name ______________________________

PRACTICE/HOMEWORK

Ratio and Proportion

Write each ratio in fraction form. Then find the unit rate.

1. 200 words in 5 min ________________

2. $45 in 5 hr ________________

3. 260 miles in 4 hr ________________

4. $20 for 8 items ________________

Solve.

5. Tell which is the better buy, a phone card for 25 minutes for $5.00 or a phone card for 60 minutes for $15.00.

____

6. A one-year magazine subscription costs $14.95. Zeta cancels the subscription after 2 months. What is her refund?

____

Tell whether the figures in each pair are similar. Write *yes* or *no*. If you write *no*, explain.

7. ____________

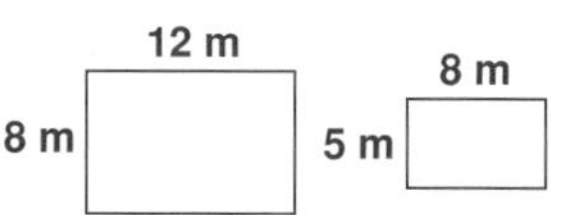

8. ____________

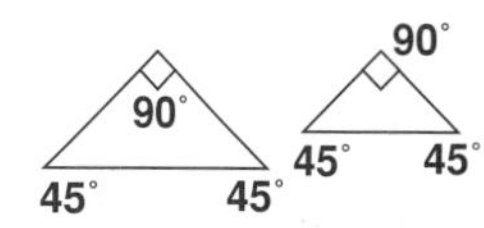

9. ____________

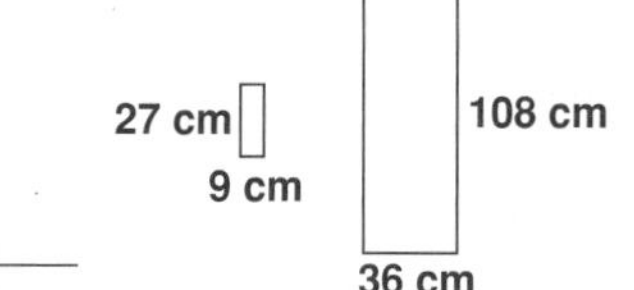

10. ____________

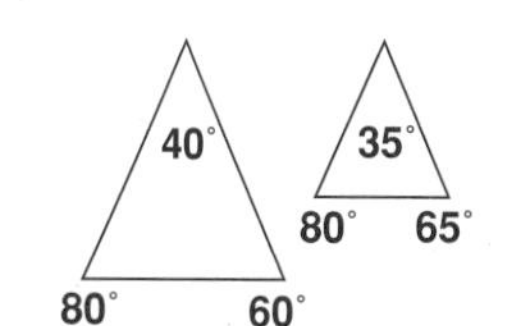

The figures are similar. Write a proportion. Then find the unknown length.

11. ________________________

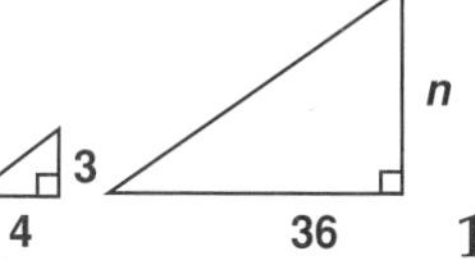

12. ________________________

n
2
18
6

Find the unknown dimension.

13. scale: 1 in. = 3.5 ft
drawing length = 8 in.
actual length = ___ ft

14. scale: 1 in. = 3.5 ft
drawing length = ___ in.
actual length = 42 ft

15. scale: 1 in. = 3.5 ft
drawing length = 11 in.
actual length = ____ ft

Use a map scale of 1 in. = 40 mi. Find the actual distance.

16. 4 in. _____

17. $\frac{3}{4}$ in. _____

18. $6\frac{1}{2}$ in. _____

19. $7\frac{1}{4}$ in. _____

Answers: 1. 200/5; 40 wpm; **2.** 45/5; $9 an hr; **3.** 260/4; 65 mph; **4.** 20/8; $2.50 each; **5.** 25 min for $5.00; **6.** $12.45; **7.** No; side length ratios are not equivalent; **8.** Yes; **9.** Yes; **10.** No; corresponding angles are not congruent; **11.** 3/4 = n/36; n = 27; **12.** 6/18 = 2/n; n = 6; **13.** 28; **14.** 12; **15.** 38.5; **16.** 160 mi; **17.** 30 mi; **18.** 260 mi; **19.** 290 mi

Family Fun

How Far Is It?

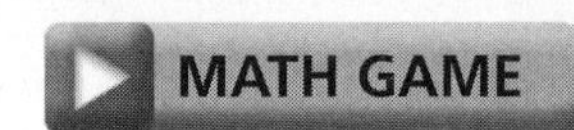

Use the world map in your social studies book or an atlas to find the distance from where you live to these cities. Use the scale and a ruler to convert inches to miles or kilometers.

How far is it:

To Nairobi, Kenya?

To Delhi, India?

To Ankara, Turkey?

To Lima, Peru?

To Beijing, China?

To Auckland, New Zealand?

HARCOURT MATH

GRADE 6

Chapter 21

WHAT WE ARE LEARNING

Percent and Change

VOCABULARY

Here are the vocabulary words we use in class:

Ratio A comparison of two numbers. Another way to express a ratio is as a percent.

Percent The ratio of a number to 100. Percent means per hundred.

Discount An amount that is subtracted from the regular price of an item

Principal The original amount put in an account. The amount the bank adds is interest.

Simple interest A fixed percent of the principal paid yearly. To calculate, use the formula $I = prt$; where I = interest earned, p = principal, r = interest rate, and t – time in years.

Name

Date

Dear Family,

Your child is learning about percent and change. Your child starts by writing ratios as percent and then relates percents, decimals, and fractions. We practice finding percents; constructing circle graphs to visualize percents; and finding discounts, sales tax, and interest.

Here are several ways to find percent:

Your teacher assigns you a 200-word essay. You write 120 words. What percent of the total have you done?

Mental Math Use mental math and friendly numbers to find the answer.

Half of 120 is 60. →
Half of 200 is 100. → $\frac{60}{100} = 60\%$

Fraction Method Write a fraction and simplify it.

$\frac{120}{200} = \frac{3}{5}$

$\frac{3}{5}$ of 100 = 60. So, it is 60%.

Proportion Method Write and solve a proportion.

$$\frac{120}{200} = \frac{n}{100}$$

$$12{,}000 = 200n \quad \text{Cross multiply.}$$

$$\frac{12{,}000}{200} = \frac{200n}{200}$$

$$n = 60\%$$

Division Method Convert the ratio to a decimal and the decimal to a percent.

$120 \div 200 = 0.6$

Multiply by 100 (move decimal point 2 places to right).

$0.6 \times 100 = 60\%$

Encourage your child to do the activity with you and other members of your family.

Sincerely,

Check It Out

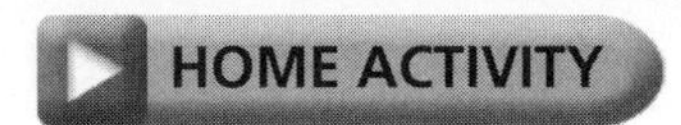

Here are some ways you can practice finding percent.

Computer

Look on your computer and find the size in megabytes of your favorite software. Compare that to the total capacity of your computer. What percent of the total capacity of your computer is the size of your favorite software?

Days Off from School

Find the total number of days you do not have school in one year. Include weekends, holidays, and vacations. Compare that to the number of days in a year. What percent of the year do you not have school?

Media

Keep track of the number of hours in one day that you listen to music or watch television. Compare that to the total hours of free time you have in one day. What percent of your free time do you listen to music or watch television?

Food Item

Find a food item you like that has a nutritional label. Compare the number of calories of your typical serving to your total caloric needs for the day (about 2400 calories for an adolescent). What percent of your daily calories is taken up by a serving of the item you found?

Your Favorite Activity

Choose an activity you enjoy. It could be related to sports, school work, music, a hobby, or any other leisure activity. How many hours a week do you spend on your activity? What percent of a week does this time represent?

Name ______________________

Percent and Change

Write the percent that is shaded.

1.

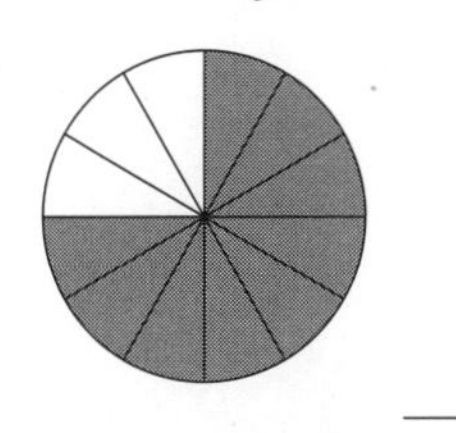

_____

2.

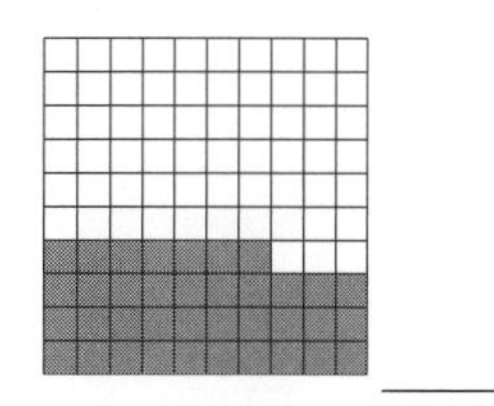

_____

3.

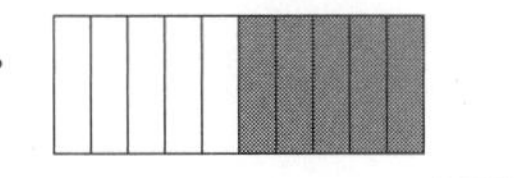

_____

4.

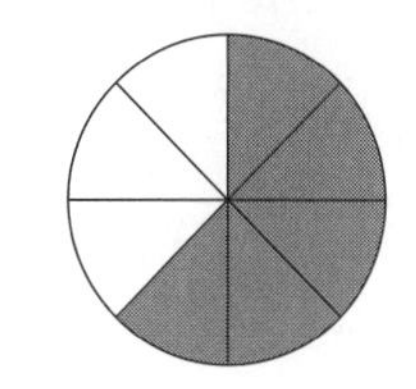

_____

Write as a percent.

5. 0.98 _____ **6.** $\frac{4}{5}$ _____ **7.** 0.04 _____ **8.** $\frac{3}{4}$ _____

Write each percent as a decimal and as a fraction in simplest form.

9. 40% _____ **10.** 94% _____ **11.** 20% _____ **12.** 55% _____

Find the percent.

13. 20% of 900 _____ **14.** 15% of 38 _____ **15.** 65% of 75 _____ **16.** 85% of 250 _____

Find the sale price.

17. regular price: $50, 15% off _____

18. regular price: $72, 30% off _____

Find the regular price.

19. sale price: $44, 45% off _____

20. sale price: $138, 25% off _____

21. Find the simple interest.

Principal	Rate	Interest/1 year	Interest/3 years
$5,500	4%		
$18,250	6.25%		

Answers: 1. 75%; **2.** 37%; **3.** 50%; **4.** 62.5%; **5.** 98%; **6.** 80%; **7.** 4%; **8.** 75%; **9.** 0.40; $\frac{2}{5}$; **10.** 0.94; $\frac{47}{50}$; **11.** 0.20; $\frac{1}{5}$; **12.** 0.55; $\frac{11}{20}$; **13.** 180; **14.** 5.7; **15.** 48.75; **16.** 212.5; **17.** $42.50; **18.** $50.40; **19.** $80; **20.** $184; **21.** $220, $660, $1,140.63, $3,421.88

Family Fun What Doe$ It Co$t?

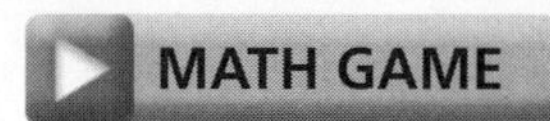

What you need:

- number cube
- game cards (See below.)
- pencil and paper

Directions:

Cut out the cards below. On each blank card, write the name of an item of interest to you and its price. Shuffle the cards and place them face down. Each roll of the number cube represents a tax rate to be used to find the total cost of an item. For example, a roll of 6 represents 6%.

Player 1 rolls the number cube and turns over the top card. The player calculates the total cost of the item including sales tax. The player records the total cost as his or her score. Player 2 goes next. Play continues until one player has spent a total of $300.

Music CD, $15.98	CD-ROM space game, $29.99
Electronic game system, $199.99	Phone card, $20.00
Boxed trading cards, $14.99	15-speed bicycle, $250.00
Leopard gecko, $30.00	

Name

Date

WHAT WE ARE LEARNING

Probability of Simple Events

VOCABULARY

Here are the vocabulary words we use in class:

Outcome Possible result of an event or experiment

Sample space The set of all possible outcomes

Theoretical probability A comparison of the number of favorable outcomes to the number of possible equally likely outcomes. It can be written as a ratio.

$$P(\text{event}) = \frac{\text{number of favorable outcomes}}{\text{number of possible equally likely outcomes}}$$

Dear Family,

Your child is working on finding the theoretical probability of an event, using a simulation to model an experiment, and on finding the experimental probability of an event.

This is how your child is learning to determine theoretical probability.

- **STEP 1**
  List the favorable outcomes.
  If you have a red marble, a clear marble, an orange marble, and a silver marble in your pocket, what is the probability that you will pull out the silver marble?
  1 favorable outcome: silver
- **STEP 2**
  Write the sample space.
  4 possible outcomes: red, clear, orange, silver
- **STEP 3**
  Write the probability as a ratio.
  $P(\text{silver}) = \frac{\text{1 favorable outcome}}{\text{4 possible outcomes}} = \frac{1}{4}$

So the probability that you will pull out the silver marble is $\frac{1}{4}$.

This is how your child is learning to determine experimental probability.

- **STEP 1**
  Understand what you are asked to find out.
  What is the probability of getting tails when you toss a coin ten times?
  You need to do an experiment and then use the results of the experiment to determine the probability.
- **STEP 2**
  Plan the strategy you will use.
  A coin toss is a good strategy.
- **STEP 3**
  Conduct the experiment and record your findings.
  Toss a coin ten times and record the number of heads and tails.
- **STEP 4**
  Determine the probability.
  If you toss tails 6 out of 10 times, then the experimental probability of tossing tails is $\frac{6}{10} = \frac{3}{5}$, or 60%.

Experimental probability The number of times a certain outcome of an event occurs compared with the total number of times you do the activity. It can be written as a ratio.

Experimental probability = $\frac{\text{number of times outcomes occurs}}{\text{total number of trials}}$

Ask questions such as these as you work together.

What prediction would you make about the number of tails you will get in ten tosses? Your child might respond: I think there will be about 5 because a coin has two sides.

What other strategy could you use? Your child may suggest: I could use a spinner with two sections, one for heads and one for tails.

Why is experimental probability useful? Your child may answer: I can use experimental probability to predict future events.

As you work with your child, talk about math to help build confidence and understanding.

Sincerely,

Name ______________________

Probability of Simple Events

Use the spinner at the right to find each probability. Express the answer as a fraction, a decimal, and a percent.

1. P(apples) ______

3. P(not bananas) ______

2. P(strawberries or apples) ______

4. P(grapes, oranges, or apples) ______

A box contains some small rings: 4 red, 6 blue, 3 white, and 7 black. You pick a ring without looking. Find each probability. Write the probability as a fraction.

5. P(red) ____ **6.** P(black or blue) ____ **7.** P(not white) ____

Alexandra throws a number cube 120 times and records her results in the table below. Write the probabilities as fractions.

8. What is the theoretical probability of throwing each number?

Number	1	2	3	4	5	6
Times Rolled	20	30	18	12	10	30

____

9. What is the experimental probability of each number?

____

10. Based on her experimental results, how many times can Alexandra expect to roll a 3 in the next 50 tosses?

____

Answers: **1.** $\frac{1}{5}$, 0.2, 20%; **2.** $\frac{2}{5}$, 0.4, 40%; **3.** $\frac{4}{5}$, 0.8, 80%; **4.** $\frac{3}{5}$, 0.6, 60%; **5.** $\frac{1}{5}$; **6.** $\frac{13}{20}$; **7.** 17/20; **8.** $\frac{1}{6}$; **9.** $\frac{1}{6}$, –, $\frac{3}{20}$, $\frac{1}{10}$, $\frac{1}{12}$, –; **10.** between 7 and 8 times

Family Fun Round and Round

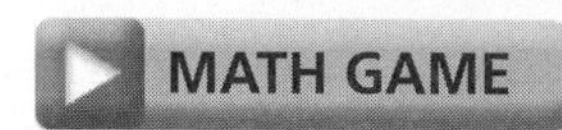

Directions:

Glue the spinners and pointers to light cardboard or construction paper and cut them out. Attach the pointers to the center of the spinners.

The objective of the game is to determine which of the three spinners is the winning spinner (lands on the greater number most often.).

You and your partner each choose and spin the pointer. Record which spinner has the greater number. Spin each pair of spinners fifty times and continue to record the results. Repeat the process so that you have statistics on using A and B, using B and C, and using A and C.

1. Which wins more often, A or B? _____
2. Which wins more often, B or C? _____
3. Which wins more often, A or C? _____
4. What is surprising about this experiment? ____________________

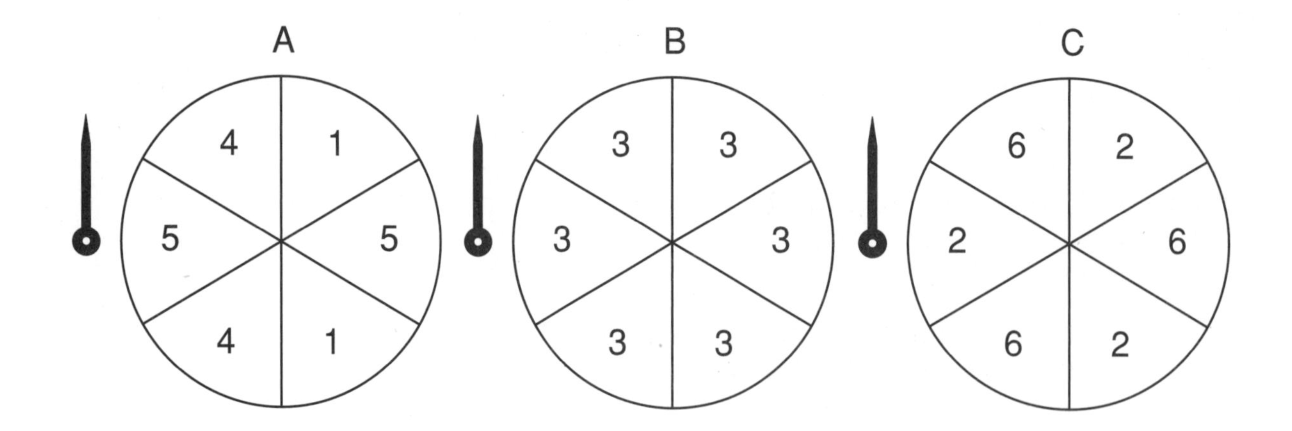

Answer: 1. A; **2.** B; **3.** C; **4.** Possible response: I thought that since A was stronger than B and B was stronger than C, A would be stronger than C

HARCOURT MATH

GRADE 6

Chapter 23

WHAT WE ARE LEARNING

Probability of Compound Events

VOCABULARY

Here are the vocabulary words we use in class:

Compound event An event that includes two or more simple events

Tree diagram A drawing to show the number of possible outcomes of a compound event

Fundamental Counting Principle If one event has m possible outcomes and a second independent event has n possible outcomes, then there are $m \times n$ total possible outcomes.

Independent events When the outcome of the second event does not depend on the outcome of the first event

Dependent events When the outcome of the second event depends on the outcome of the first event

Name

Date

Dear Family,

Your child is learning about the probability of compound events. Your child will learn to find outcomes of compound events, identify and find probabilities of independent and dependent events, and find probabilities and make predictions from sample data.

Here is how your child is learning to find the probability of independent and dependent events.

A box holds cards labeled 1–6. Without looking, you pull out a card labeled 4 and replace it. You pull out another card labeled 1. What is the probability of selecting a 4 and then a 1?

- **STEP 1**
  Find the probability of the first selection.
  $P(4) = \frac{1}{6}$
- **STEP 2**
  Find the probability of the second selection.
  $P(1) = \frac{1}{6}$
- **STEP 3**
  Multiply the probabilities.
  $\frac{1}{6} \times \frac{1}{6} = \frac{1}{36}$ So, the probability is $\frac{1}{36}$.

Now, suppose you do not replace the first card before pulling out a second card. What is the probability of selecting a 4 and then a 1?

- **STEP 1**
  Find the probability of the first selection.
  $P(4) = \frac{1}{6}$
- **STEP 2**
  Find the probability of the second selection.
  $P(1) = \frac{1}{5}$
- **STEP 3**
  Multiply the probabilities.
  $\frac{1}{6} \times \frac{1}{5} = \frac{1}{30}$ So, the probability is $\frac{1}{30}$.

Use the model here and the practice activities on the back of this page to help your child practice finding probabilities of independent and dependent events.

Sincerely,

Independent and Dependent Events

Decide if these events are independent or dependent. Find the probabilities.

1. You toss two coins. What is the probability of getting the following:

   a. two heads? ______________________

   b. two tails? ______________________

   c. one head and one tail? ______________________

2. You toss one coin and roll a number cube labeled 1–6. What is the possibility of getting the following:

   a. heads and the number 5? ______________________

   b. tails and the number 7? ______________________

3. Ten cards labeled 1–10 are in an envelope. At random, you select a card, and it is a 9. You lay it on the table and select another card, and it is a 2. What is the probability of selecting a 9 and then a 2?

You have a bag with 7 marbles – 2 red, 2 blue, 3 white.

4. You pull out 1 marble from a bag, return it to the bag, and pull out another marble. What is the probability that you pull out a blue marble each time?

   ____

5. You pull out 1 marble from a bag, return it to the bag, and pull out another marble. What is the probability that you pull out a blue marble and then a white marble?

   ____

6. You pull out 1 marble from a bag, do not return it to the bag, and pull out another marble. What is the probability that you pull out two blue marbles?

   ____

Answers: 1. independent; $\frac{1}{4}$; $\frac{1}{4}$; $\frac{2}{4}$ or $\frac{1}{2}$; **2.** independent; $\frac{1}{12}$, 0; **3.** dependent; $\frac{1}{90}$; **4.** independent; $\frac{4}{49}$; **5.** independent; $\frac{6}{49}$; **6.** dependent; $\frac{1}{21}$

Name ______________________________

PRACTICE/HOMEWORK

Probability of Compound Events

Draw a tree diagram or make a table to find the number of possible outcomes for each situation.

1. Derek can do his math homework with a wooden pencil, a mechanical pencil, or an erasable pen. He can use graph paper or lined paper. _____

2. Emily can order a blueberry, cinnamon, plain, onion, or sesame bagel. Her choice of spreads is cream cheese, butter, jam, or humus. _____

Use the Fundamental Counting Principle to find the number of possible outcomes for each situation.

3. A choice of 3 solid-colored shorts, 5 striped shirts, and 3 styles of shoes.

____

4. A choice of 3 ice cream flavors, 7 candy toppings, and whipped cream or fudge sauce.

____

For 5–8, use the spinner and cards. Suppose you spin the pointer on the spinner and select a card at random.

5. Find P(whale, poster). _____
7. Find P(stingray, model). _____
6. P(dolphin, speech.) _____
8. P(shark, speech or poster). _____

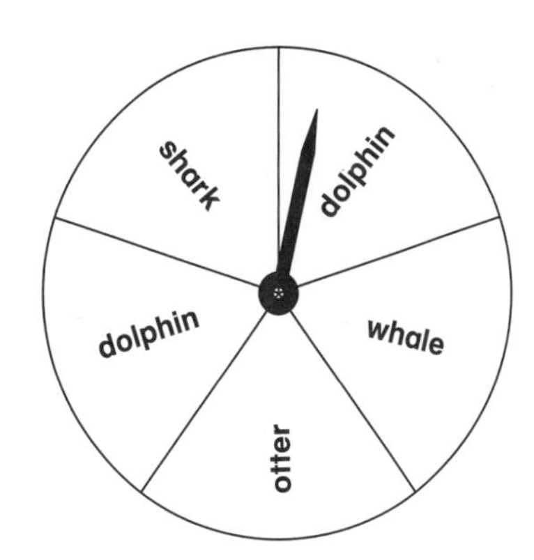

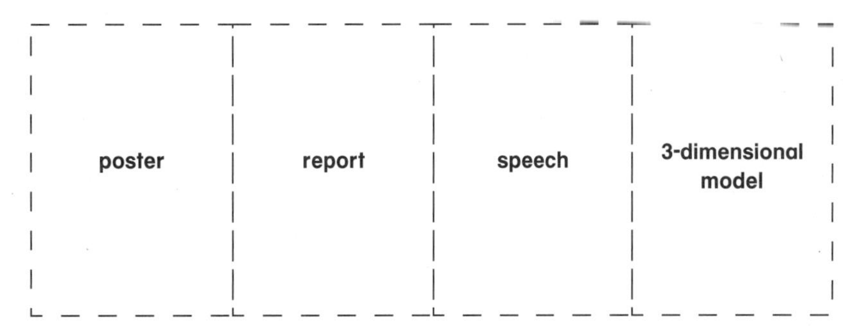

Answers: For **1 and 2,** check students' diagrams or tables. **1.** 6 outcomes; **2.** 20 outcomes; **3.** 3 x 5 x 3 = 45, 45 outcomes; **4.** 3 x 7 x 2 = 42, 42 outcomes; **5.** $\frac{1}{20}$; **6.** $\frac{1}{10}$; **7.** 0; **8.** $\frac{1}{10}$

Family Fun

MATH GAME

You have eight letter tiles labeled E, F, N, O, O, O, T, and T. You turn them facedown on a table, mix them, and then turn over one tile at a time. Surprise, the tiles make a word in the order you turned them over. You spelled FOOTNOTE. Just like that!

Find the probability of turning over the first tile and finding F, the second and third tiles and finding O, the fourth tile and finding T, and so on to spell FOOTNOTE.

What is the probability of this happening?

Answers: $\frac{1}{8} \times \frac{3}{7} \times \frac{2}{6} \times \frac{2}{5} \times \frac{1}{4} \times \frac{1}{3} \times \frac{1}{2} \times \frac{1}{1} = \frac{12}{40{,}320}$ or 1:3,360

HARCOURT MATH

GRADE 6

Chapter 24

WHAT WE ARE LEARNING

Units of Measure

VOCABULARY

Here are the vocabulary words we use in class:

Precision A property of measurement that is related to the unit of measure used; the smaller the unit of measure used, the more precise the measurement is.

Name

Date

Dear Family,

Your child is studying customary and metric measurement. This is how your child is learning to use and relate customary and metric measurements.

- **STEP 1**
  Use the relationship between the measurements to write a proportion to change one unit to another.

$$\frac{1 \text{ ft}}{12 \text{ in.}} = \frac{3 \text{ ft}}{x \text{ in.}} \qquad \frac{1 \text{ ft}}{30.48 \text{ cm}} = \frac{3 \text{ ft}}{x \text{ cm}}$$

- **STEP 2**
  Find the cross products.

$x = 3 \times 12$
$x = 36$

$x = 3 \times 30.48$
$x = 91.44$

Ask questions such as these as you work together.

How do you find the number of feet in 5 meters? Your child might begin: First I use the relationship of meters to feet to write a proportion. It is $\frac{1 \text{ m}}{3.28 \text{ ft}} = \frac{5 \text{ m}}{x \text{ ft}}$

What is the next step? Your child might say: I find the cross products and multiply, so $x = 3.28 \times 5$.

Then, what do you do? Your child might respond: I solve for x, so $x = 16.4$.

Tell me your answer. Your child might reply: There are 16.4 feet in 5 meters.

This is how your child is learning to use the appropriate mathematical tools and units of measure to solve problems.

Recognize that measurements in customary and metric units for length, capacity, temperature, and weight (or mass) are approximations.

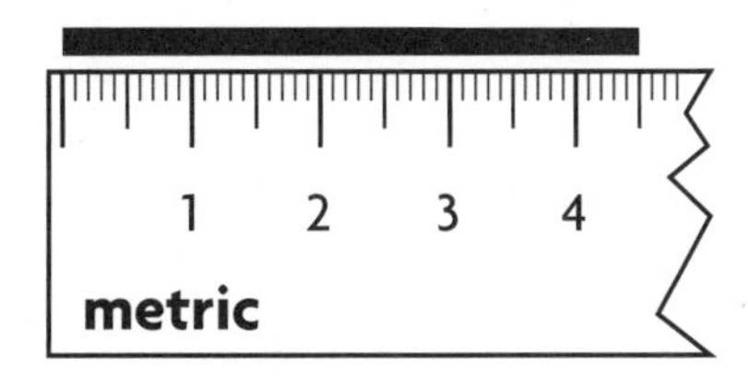

4.5 cm or 45 mm

Understand that the smaller the unit of measure used, the more precise the measurements.

To measure an object, decide which unit of measure is most appropriate; generally, the smaller the object to be measured, the smaller the unit used to measure.

To weigh fruit, use kilograms, not grams.

To measure a pencil tip, use millimeters, not centimeters.

Why is the mass of a workbook more precise if it is measured to the nearest gram, not kilogram? Your child might reply: A gram measure is more precise because it is a smaller unit than a kilogram.

What unit would you use to measure the length of a driveway? Your child might respond: I would use meters or yards.

What unit would you use to measure juice in a drinking glass? Your child might reply: I would use ounces or milliliters.

As you work with your child, talk about math to help build confidence and understanding.

Sincerely,

Name ______________________

Units of Measure

Convert these customary measures to the given unit by using a proportion.

1. 18 ft = ____ yd
2. 8 pt = ____ qt
3. 6 in. = ____ yd

Convert these metric measures to the given unit by using a proportion.

4. 3.7 m = ____ km
5. 0.25 L = ____ mL
6. 2,000 cm = ____ m

Convert these customary and metric measures to the given unit by using a proportion.

7. 22 yd ≈ ____ m
8. 5 L ≈ ____ gal
9. 12.7 cm ≈ ____ in.

Measure the line segment to the given length.

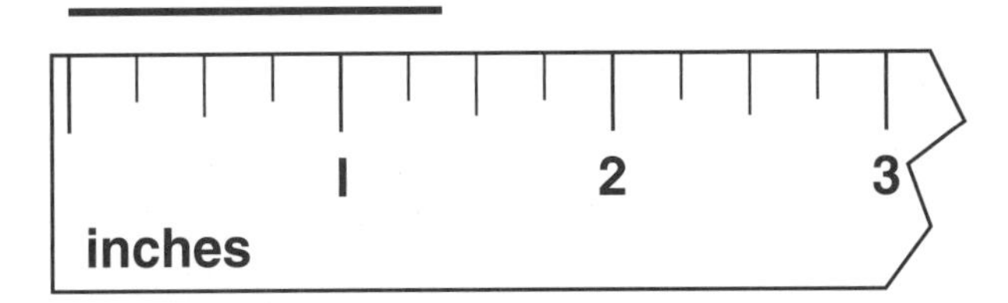

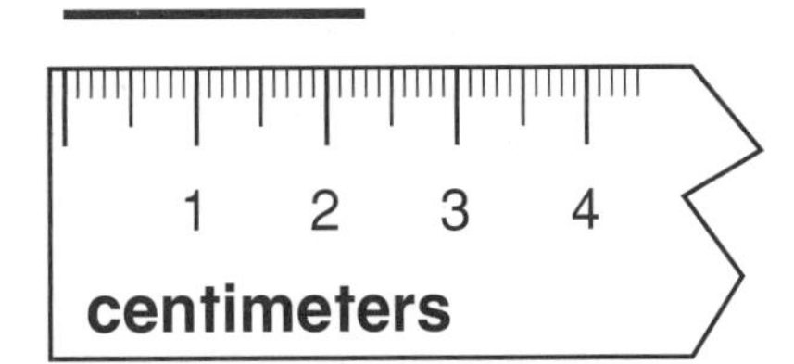

10. nearest half inch ____
11. nearest inch ____
12. nearest centimeter ____
13. nearest millimeter ____

Tell which measurement is more precise.

14. 3 qt or 13 c ____
15. 1,490 m or 1 km ____
16. 5 oz or $\frac{1}{2}$ c ____

Name the most appropriate unit of measure for each item.

17. Water in a swimming pool

____

18. Weight (mass) of a bag of flour

____

Answers: **1.** 6; **2.** 4; **3.** $\frac{1}{6}$; **4.** 0.0037; **5.** 250; **6.** 20; **7.** 20.02; **8.** 1.32; **9.** 5; **10.** $1\frac{1}{2}$ in.; **11.** 1 in.; **12.** 2 cm; **13.** 23 mm; **14.** 13 c; **15.** 1,490 m; **16.** 5oz; **17.** gallons or liters; **18.** pounds or kilograms

Family Fun

A CROSS-NUMBER PUZZLE

MATH GAME

Directions:
Omit decimal points and percent signs in the puzzle. Write only the digits.

1	2	3			4	5	
	6			7			
	8		9			10	11
12			13	14			
		15				16	
17					18		19
		20		21			
22				23			

Across

1. How many yards are in 4,890 feet?
4. How many inches are in 19 feet?
6. 8 dozen
7. Five minutes past eight
8. How many acres are in one square mile?
10. XLIV
13. What is the total cost of $4\frac{3}{4}$ yard of plastic sheeting at $0.40 per yard, one roll of tape at $0.10, and 3 packages of tacks at $0.05 each?
15. What is the mean of 728, 964, 247, 425, and 316?
17. What is 90% of 100?
18. How many feet are there in 104 yards?
21. What is the cost of 1 bag of crackers if 27 bags costs $11.61?
22. How many quarters are there in $33.75?
23. Write 44% as a decimal.

Down

2. To the nearest gram, how many grams are there in 24.56 ounces?
3. What is 1 less than the number of days in a year that is not a leap year?
4. Express $\frac{1}{5}$ as a percent.
5. How much interest will $127.00 earn in one year if the rate is 2%?
9. Express $\frac{1}{4}$ mile in feet. Then reverse the digits.
11. To the nearest meter, how many meters are there in 160 feet?
12. What is the total cost, including 2% tax, for 1 dozen individual bags of peanuts at $0.06 each, 2 packages of cheese and crackers at $0.17 each, and 1 carton of milk at $0.40?
14. How many ounces are in a pound?
16. How many dozen energy bars are needed to feed the 33 players on the football team if each player gets 4 bars?
18. How many days remain in a year that is not a leap year after January has passed?
19. The age in 2007 of a woman born in 1980
20. Write $\frac{1}{4}$ as a decimal.
21. How many pounds are there in 20 kilograms?

Answers: 1. 1,630; **4.** 228; **6.** 96; **7.** 805; **8.** 640; **10.** 44; **13.** 215; **15.** 536; **17.** 90; **18.** 312; **21.** 43; **22.** 135; **23.** 44; **Down: 2.** 696; **3.** 364; **4.** 20; **5.** 254; **9.** 0231; **11.** 49; **12.** 149; **14.** 16; **16.** 11; **18.** 334; **19.** 27; **20.** 25; **21.** 44

HARCOURT MATH

GRADE 6

Chapter 25

WHAT WE ARE LEARNING

Length and Perimeter

VOCABULARY

Here are the vocabulary words we use in class:

Circumference The distance around a circle

Pi (π) The ratio of the circumference to the diameter; $\pi \approx 3.14$ or $\frac{22}{7}$

Name

Date

Dear Family,

Your child is studying perimeter and circumference of geometric shapes. This is how your child is learning to find the perimeter of polygons.

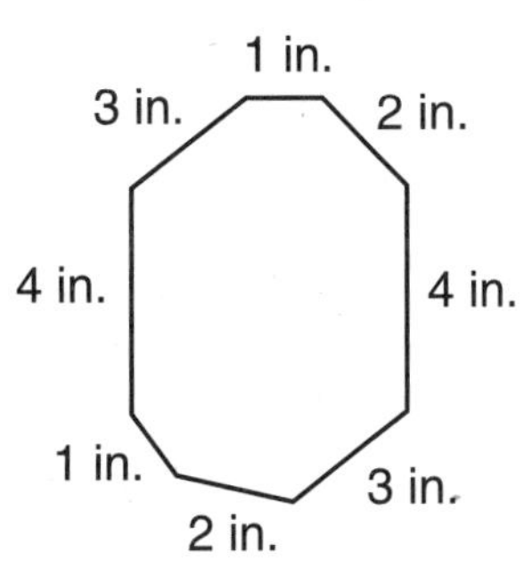

- **STEP 1**
  The perimeter of any polygon can be found by adding the lengths of all the sides. For the octagon to the left, you can use the formula,
  $P = a + b + c + d + e + f + g + h$
- **STEP 2**
  Replace the variables with the number of inches in the lengths of the sides.
  $P = a + b + c + d + e + f + g + h$
  $P = 1 + 2 + 4 + 3 + 2 + 1 + 4 + 3$
- **STEP 3**
  Solve the equation.
  $P = 20$ So, the perimeter is 20 inches.

Ask questions such as these as you work together.

How would you find the perimeter of a rectangle with a width of 2 inches and a length of 12 inches? Your child might respond: The formula for the perimeter of a rectangle is $P = 2l + 2w$.

What is the next step? Your child might explain: I substitute known values into the equation.
$P = 2 \times 12 + 2 \times 2$

What do you do next? Your child might reply: I simplify the expression by adding the products. $P = 24 + 4 = 28$ in. The perimeter of the rectangle is 28 inches.

This is how your child is learning to find the circumference of a circle.

- **STEP 1**
  Use the formula for finding the circumference of a circle.
  $C = \pi d$

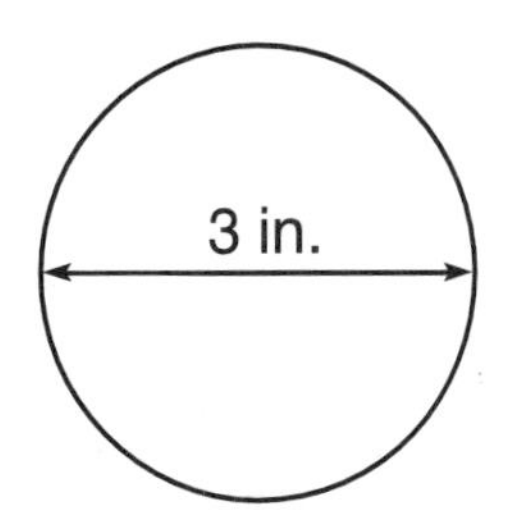

- **STEP 2**
  Replace pi with 3.14 and substitute the known value for d.
  $C \approx 3.14 \times 3$
- **STEP 3**
  Solve the equation.
  $C \approx 9.42$

Show me the steps you use to find the circumference of a circle with a 2-inch radius. Your child might begin: I must write the formula for this circle, $C = \pi d$. since a radius is half of the diameter, I double the 2 inches and the diameter is 4 inches.

What do you do next? Your child might respond: I substitute the known values into the formula, so $C \approx 3.14 \times 4$.

What is your final step? I solve the equation, so $C \approx 12.56$.

As you work with your child, talk about math to help build confidence and understanding.

Sincerely,

Name ______________________________ PRACTICE/HOMEWORK

Length and Perimeter

Find the perimeter.

1. ________ 3 ft, 8 ft, 5 ft

2. ________ 5 cm, 10 cm, 10 cm, 5 cm

3. ________ 5.5 yd, 5.5 yd, 5.5 yd, 5.5 yd, 5.5 yd, 5.5 yd

Find the unknown length. The perimeter is given.

4. ________ P = 37

15, 15, x

5. ________ P = 177

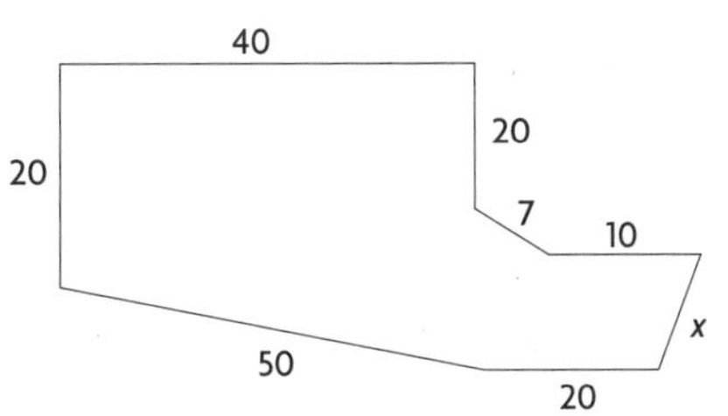

6. ________ P = 27

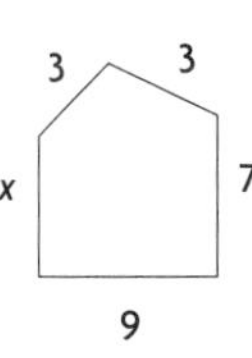

Write an equation for finding the unknown side. The perimeter is given.

7. ________ P = 28

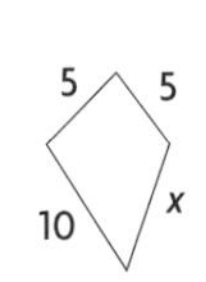

8. ________ P = 300

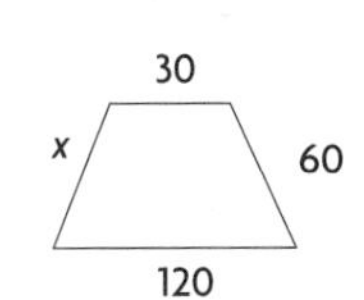

9. ________ P = 89

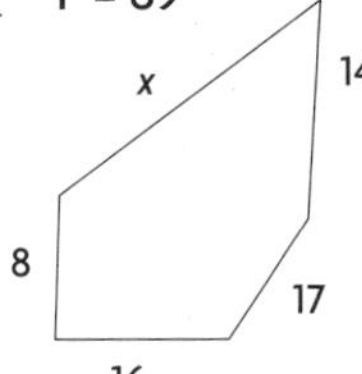

Find the circumference. Use the formula $C = \pi d$. Use 3.14 for π. Round to the nearest tenth.

10. ________

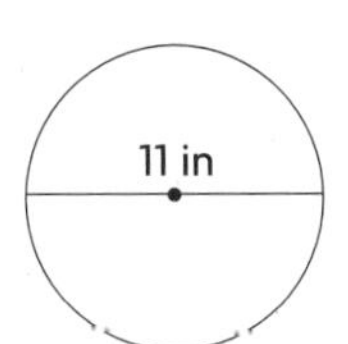

11. ________

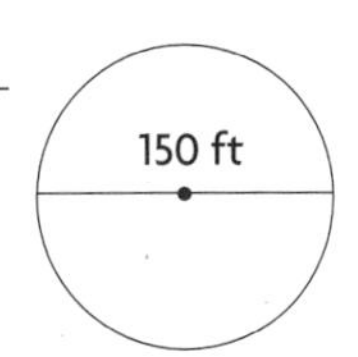

12. ________

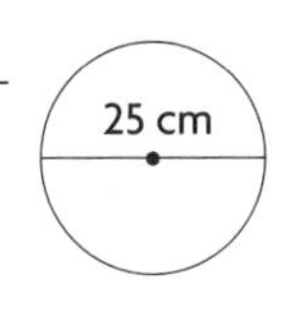

Find the circumference. Use the formula $C = 2\pi r$. Use 3.14 for π. Round to the nearest tenth.

13. ________

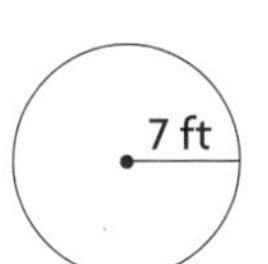

14. ________

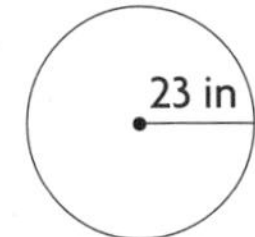

15. ________

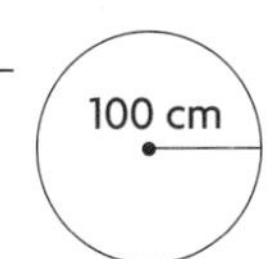

Answers: 1. 16 ft; 2. 30 cm; 3. 33 yd; 4. 7; 5. 10; 6. 5; 7. $28 = 5 + 5 + 10 + x$; 8. $300 = 30 + 60 + 120 + x$; 9. $89 = 14 + 17 + 16 + 8 + x$; 10. 34.5 in.; 11. 471 ft; 12. 78.5 cm; 13. 44 ft; 14. 144.4 in.; 15. 628 cm

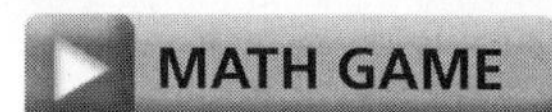

Family Fun

Directions:

Draw the geometric figures described below. Cut them out and arrange them to make a geometric figure or design. Use all the shapes. Color your results, if you wish.

1. A triangle, 1 inch on each side
2. A rectangle, $1\frac{1}{2}$ inches by 1 inch
3. A square, 1 inch on each side
4. A triangle, $1\frac{1}{2}$ inches on each side
5. A circle, 2 inches in diameter
6. A rectangle, $1\frac{1}{2}$ inches long and $\frac{3}{4}$ inch wide
7. A circle, with a diameter of $1\frac{1}{4}$ inches
8. A rhombus, 1 inch on each side
9. A rectangle, $1\frac{1}{4}$ inches by $\frac{3}{4}$ inch
10. A hexagon, $\frac{3}{4}$ inch on each side

HARCOURT MATH

Name

GRADE 6

Date

Chapter 26

WHAT WE ARE LEARNING

Area

VOCABULARY

Here are the vocabulary words we use in class:

Area The area of a plane figure is the number of square units needed to cover it. Area is measured in square units such as square feet (ft^2) or square meters (m^2).

Surface area The sum of the areas of the faces of a solid figure.

Here are the formulas used in this chapter to find **area**.

Rectangle $A = lw$

Square $A = s^2$

Triangle $A = \frac{1}{2}bh$

Parallelogram $A = bh$

Trapezoid

$A = \frac{1}{2}h(b_1 + b_2)$

Circle $A = \pi r^2$

Dear Family,

Your child is learning to find the area of plane figures and the surface area of solid figures. We estimate the area of a figure and then find the actual area by writing a formula, replacing the variables with numbers, and computing. We find the surface areas of prisms and pyramids by using a net and a formula.

Here is how we find the surface area of a prism:
The surface area is the sum of the areas of the faces of a solid figure.

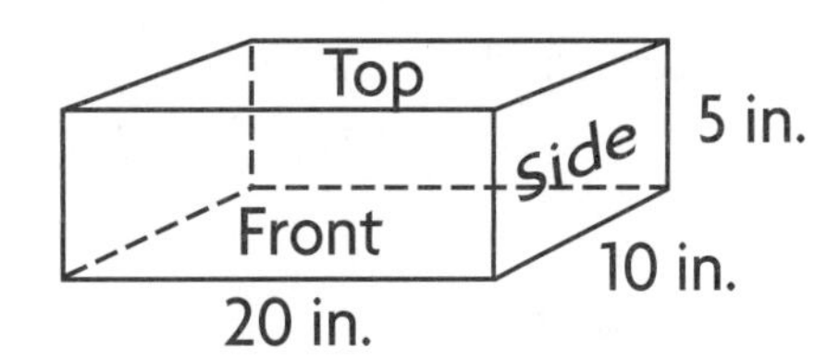

- **STEP 1**
  Use the formula for the area of a rectangle to find the area of each face.
  front and back: $(20 \times 5) \times 2 = 200$
  top and bottom: $(20 \times 10) \times 2 = 400$
  left and right sides: $(10 \times 5) \times 2 = 100$

- **STEP 2**
  Find the sum.
  $200 + 400 + 100 = 700$
  So, the surface area is 700 in^2.

Use the activities on the back of this page to help your child practice finding the surface area of prisms.

The ability to estimate and measure area is useful in real-life activities such as buying paint, fabric, carpet, tile, and art supplies.

Sincerely,

The Box Tower

Find 5 boxes in your house to measure.
Write the length, width, and height of each box.
Find the surface area of each box.

Here are some sample dimensions:

cereal box	11 in. x 7.5 in. x 2.5 in.
rice box	6.5 in. x 4 in. x 1 in.
tissue box	9.5 in. x 5 in. x 3 in.
shoe box	13 in. x 5 in. x 4 in.
box of teabags	5.5 in. x 3 in. x 3 in.

Is the sum of the surface areas for the five boxes you found greater than or less than the sum of the surface areas for the five boxes listed above? Explain how you know.

____

____

____

____

____

____

____

____

Name ______________________

PRACTICE/HOMEWORK

Area

Estimate the area. Each square on the grid represents 1 cm^2.

1. ______

2. ______

Find the area.

3. 19 cm, 14 cm ______

4. 16 cm, 14 cm ______

5. 4 ft, 5 ft ______

Find the area of each circle. Round to the nearest tenth. Use 3.14 for π.

6. 18 cm ______

7. 8 cm ______

8. $r = 5$ m ______

9. $d = 24$ mm ______

Find the surface area.

10. 6 m, 3 m, 3 m ______

11. 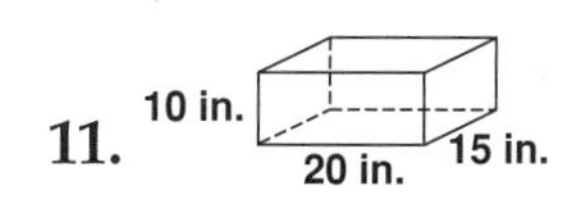

______

12. 2 m, 3 m, 2 m ______

13. What is the area of a semicircle with a radius of 5 cm? ______

14. The area of the base of a square pyramid is 144 cm^2. Each triangular face has a height of 8 cm. What is the surface area of the pyramid?

____

Draw the net.

Answers: 1. about 16 cm^2; 2. about 20 cm^2; 3. 266 cm^2; 4. 112 cm^2; 5. 10 ft^2; 6. 254.3 cm^2; 7. 201.0 in^2; 8. 78.5 m^2; 9. 452.2 mm^2; 10. 45 m^2; 11. 1200 in^2; 12. 32 m^2; 13. 39.25 cm^2; 14. 336 cm^2

Family Fun

WHO HAS MOST?

MATH GAME

What you need:

- paper
- pencils
- game cards (See below.)

Directions:

1. Cut out the cards, shuffle them, and place them face down.
2. Each player takes a card and calculates the area of the figure on the card.
3. Players check each other's calculations. If the calculation is correct, the player scores 10 points. If the calculation is incorrect, the player loses 5 points.
4. The player with the most points after all the cards have been used is the winner.

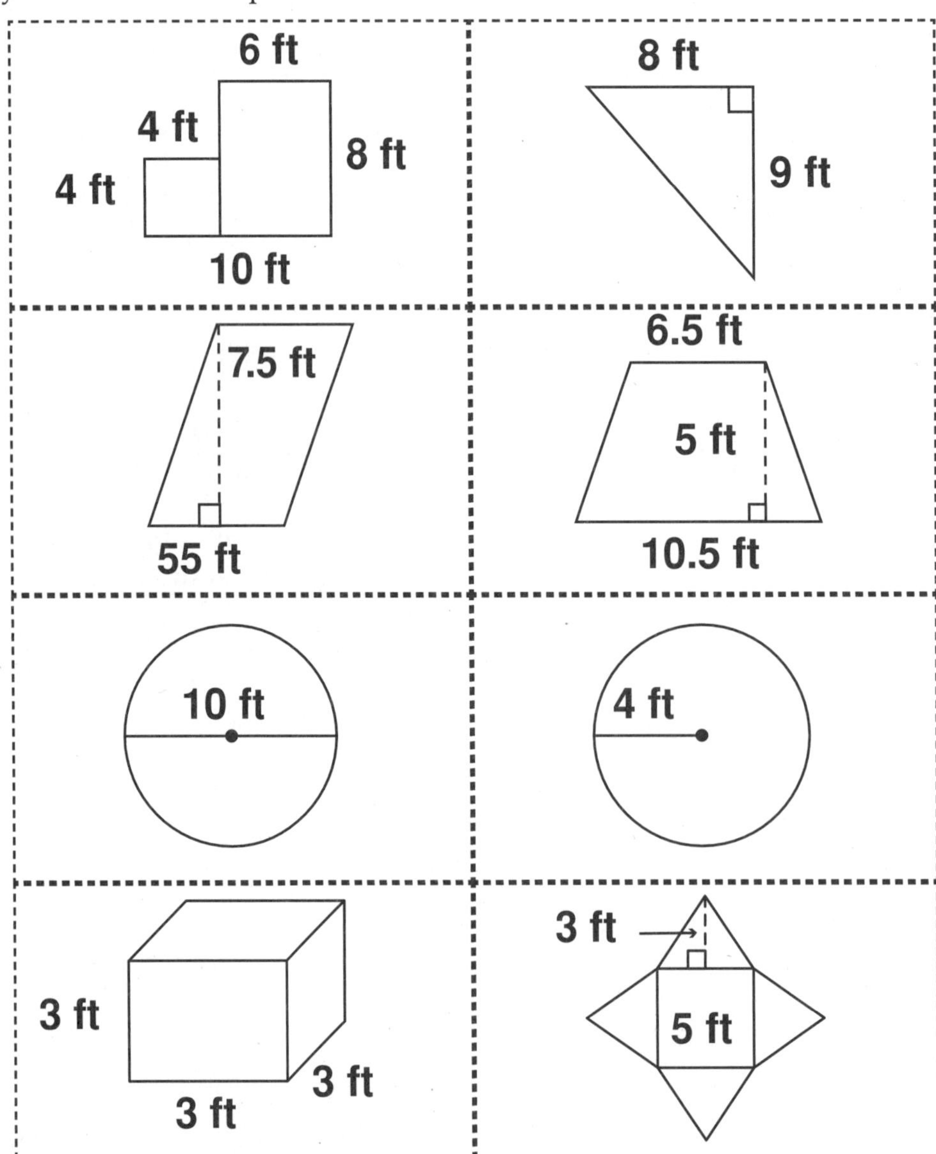

HARCOURT MATH

GRADE 6

Chapter 27

WHAT WE ARE LEARNING

Volume

VOCABULARY

Here are the vocabulary words we use in class:

Volume The number of cubic units needed to occupy a given space

Here are the formulas we use to find volume:

Volume of Prisms

Rectangular prism $V = Bh$, where B is the area of the base

Triangular prism

$V = \frac{1}{2}lwh$ or $V = Bh$

Volume of Pyramids

Square or rectangular base

$V = \frac{1}{3}Bh$

Volume of Cylinders

$V = \pi r^2 h$, where πr^2 is the area of the base, which is a circle

Name

Date

Dear Family,

Your child is learning to estimate and find the volume of prisms, pyramids, and cylinders. We find volume by writing a formula, replacing the variables with numbers, and multiplying.

Find the volume of the cylinder.

diameter = 9 cm
height = 3.5 cm

- **STEP 1**
  Find the radius.
  $9 \div 2 = 4.5$ cm

- **STEP 2**
  Write the formula
  $V = \pi r^2 h$

- **STEP 3**
  Replace π with 3.14 or $\frac{22}{7}$, r with 4.5, and h with 3.5.
  $V \approx 3.14 \times (4.5)^2 \times 3.5$

- **STEP 4**
  Multiply.
  $V \approx 3.14 \times 20.25 \times 3.5$
  $V \approx 222.5475$

So, the volume is about 223 cm^3.

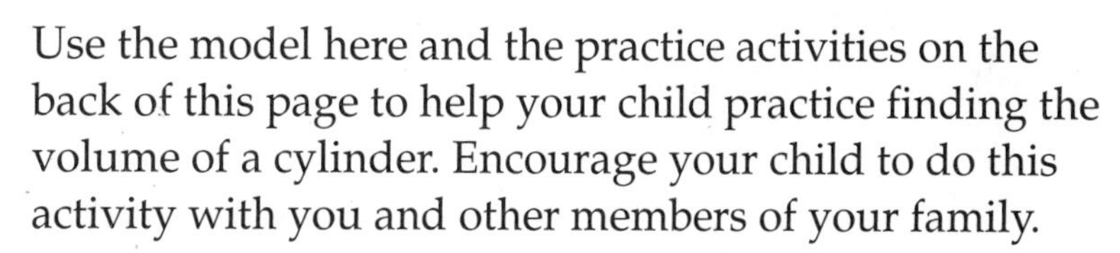

Use the model here and the practice activities on the back of this page to help your child practice finding the volume of a cylinder. Encourage your child to do this activity with you and other members of your family.

Sincerely,

Volume of Cylinders

1. Find 5 cylinders in your house to measure. Write down the diameter and height of each. Find the volume of each cylinder. Use 3.14 for π.

Complete the chart. Use 3.14 for π.

	Container	diameter	radius	height	volume
2.	can of pumpkin	4 in.		4.5 in.	
3.	can of tomato paste	6 cm		8 cm	
4.	jar of ground ginger	10 in.		2 in.	
5.	can of soup	6.4 cm		10 cm	

6. Find the volume of a roll of paper towels. The total radius is 6 cm. The radius of the inner roll is 2 cm. The height of the roll is 28 cm. Use 3.14 for π.

Answers: 1. Check students' work; **2.** 2 in., 56.52 in.^{3}; **3.** 3 cm, 226.08 cm^3; **4.** 5 in., 157 in.^{3}; **5.** 3.2 cm, 321.536 cm^3;
6. 2,813.44 cm^3

Name ______________________

Volume

Find the volume.

1. 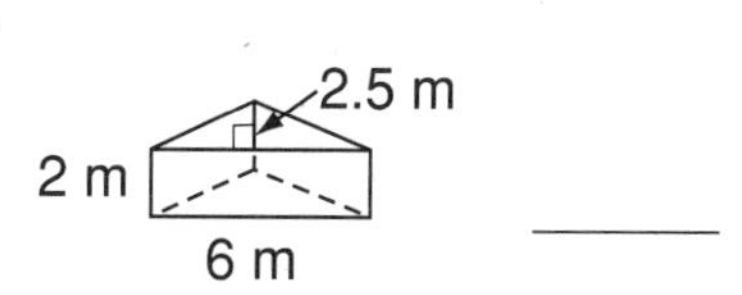

______

2. 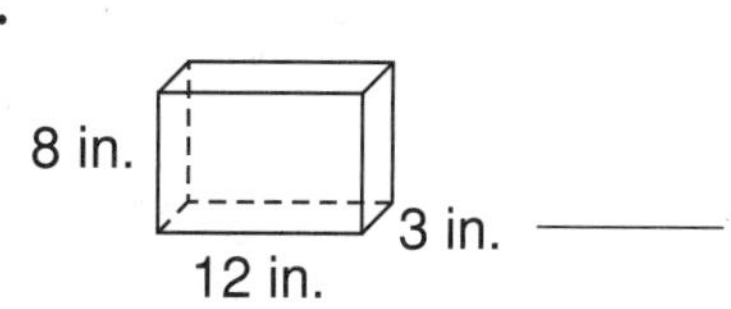

______

3. 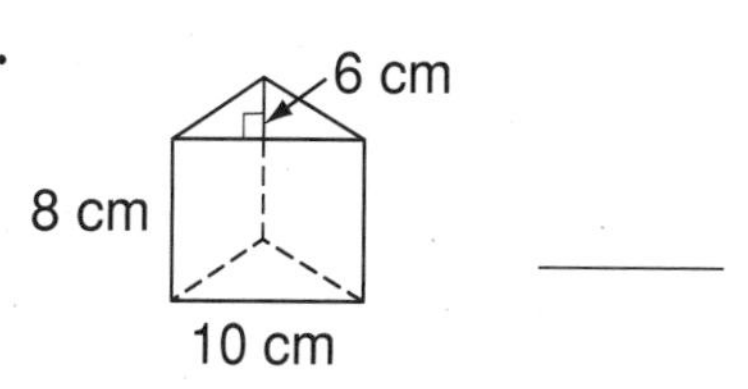

______

4. 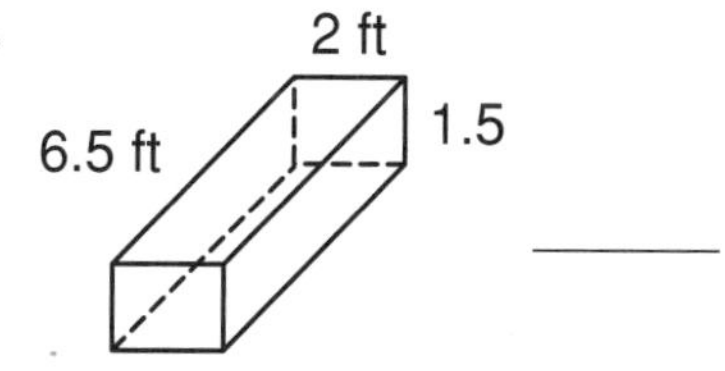

______

5. 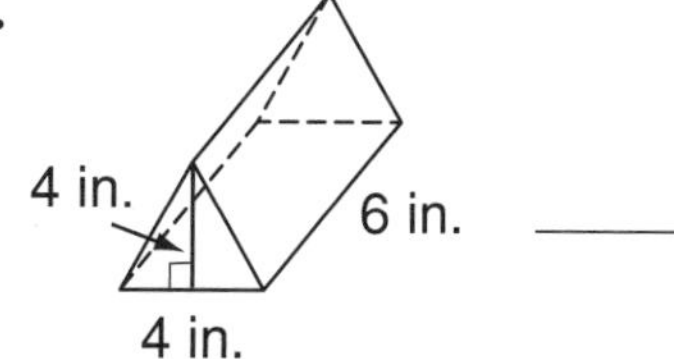

______

6. 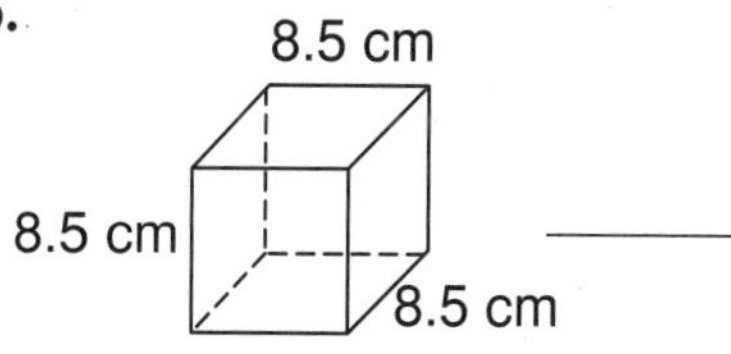

______

7. 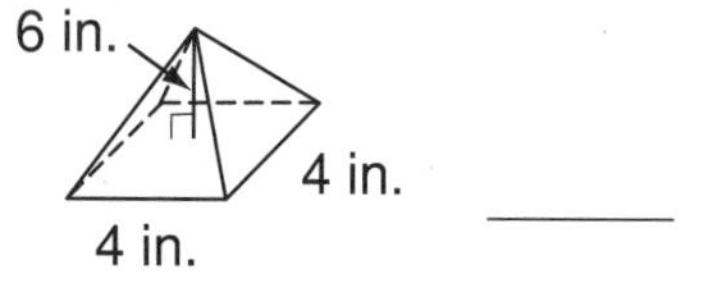

______

8. 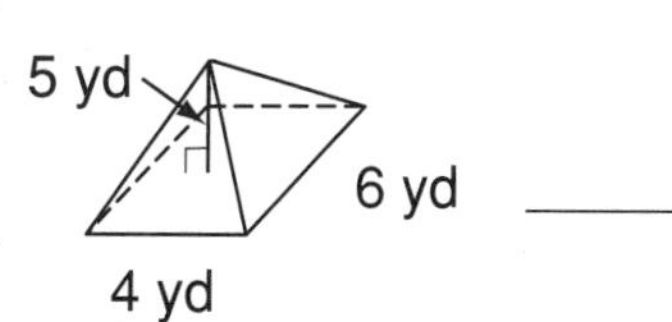

______

9. 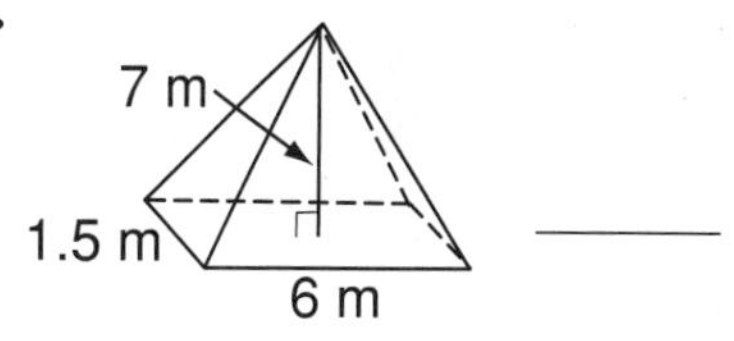

______

10. Find the volume of a rectangular pyramid with a length of 15 in., width of 12 in., and height of 20 in. ______

11. Find the volume of a square pyramid with a base of area 144 cm^2 and a height of 10 cm. ______

Find the volume. Round to the nearest whole number. Use 3.14 for π.

12. 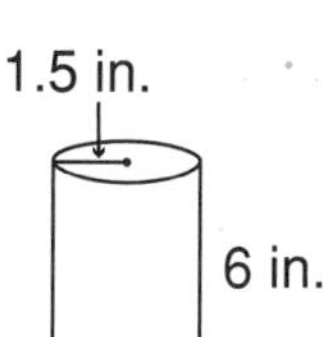

______

13. 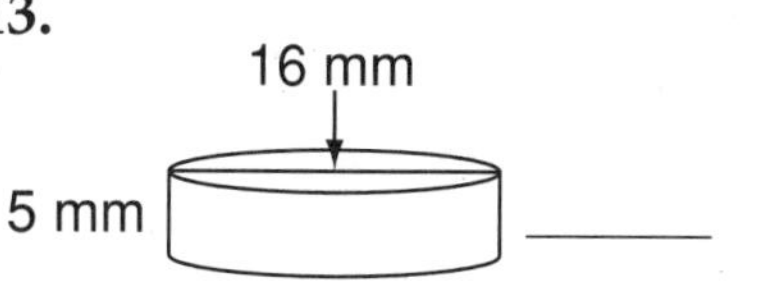

______

14. 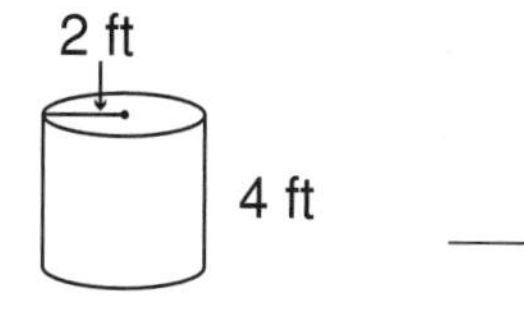

______

Answers: 1. 15 m^3; **2.** 288 $in.^3$; **3.** 240 cm^3; **4.** 19.5 ft^3; **5.** 48 $in.^3$; **6.** 614.125 cm^3; **7.** 32 $in.^3$; **8.** 40 yd^3; **9.** 21 m^3; **10.** 1,200 $in.^3$; **11.** 480 cm^3; **12.** 42 $in.^3$; **13.** 1,005 mm^3; **14.** 50 ft^3

MATH GAME

Materials

- oatmeal container
- light colored construction paper
- tape
- crayons or markers
- scissors

Directions

Work with a partner. Measure the height and diameter of an oatmeal container. Find the volume. Design and build a new container that has the same volume but has a different shape.

Name

Date

WHAT WE ARE LEARNING

Patterns

VOCABULARY

Here are the vocabulary words we use in class:

Triangular number A number that can be represented by a triangular array

Sequence An ordered set of numbers

Term Each number in a sequence

Function A relationship between two quantities in which one quantity depends on the other

Fractals A repeating pattern of smaller shapes appearing in geometric figures

Iteration A step in the process of repeating something over and over again

Dear Family,

Your child is learning about patterns. Your child will identify, extend, and make number patterns in sequences, write a rule for the function, and also identify, extend, and make patterns with geometric figures.

Here is how we identify a function and write an equation to represent it. Find the missing term in the table.

x	9	18	27	36	45
y	19	37	55	73	■

- **STEP 1**
  Look for a pattern.
  As x increases by 9, y increases by 18.
- **STEP 2**
  Write an equation that expresses one term in relation to the other.
  Think: How is each x-value related to the y-value?
  y is 1 more than 2 times x.
  $y = 2x + 1$
- **STEP 3**
  Use the equation to find the missing term.
  Write the equation: $y = 2x + 1$
  Replace x with 45: $y = 2(45) + 1$
  $y = 91$
  So, the missing term is 91.

Use the model here and the practice activities on the back of this page to help your child practice identifying a function and representing it with an equation. Encourage your child to do this activity with you and other members of your family.

Sincerely,

Writing an Equation

Identify the function in each table and represent it with an equation. Find the missing term.

1.

x	23	24	25	26	27
y	18	19	20	21	

____

2.

x	56	66	76	86	96
y	28	33	38		48

____

3.

x	5	8	10	20	25
y	17	26	32	62	

____

Write an equation to solve these problems. You may wish to use a table as part of the process.

4. Dan drove 350 miles to Santa Barbara in 5 hours. How many miles had he gone after 3 hours?

time	1	2	3	4	5
distance					350

5. Jenny bought a CD for \$16.00. The tax rate was 5%. What was her total cost?

amount	\$1	\$2	\$3	\$5	\$10	\$16
tax	\$0.05	\$0.10	\$0.15			

Answers: 1. $y = x - 5$; ? = 22; **2.** $y = x \div 2$; ? = 43; **3.** $y = 3x + 2$; ? = 77; **4.** 210 miles; **5.** \$16.80

Name ____________________ **PRACTICE/HOMEWORK**

Patterns

Write the rule for each sequence.

1. 135, 120, 105, 90, ... ________
2. 0.6, 0.36, 0.216, 0.1296, ... ________
3. $\frac{1}{6}, \frac{1}{3}, \frac{1}{2}, \frac{2}{3}, \ldots$ ________
4. 36, 9, 2.25, 0.56, ... ________

Find the next three terms in each sequence.

5. 25, 32, 39, 46, ... ________
6. $\frac{7}{8}, \frac{3}{4}, \frac{5}{8}, \frac{1}{2}, \ldots$ ________
7. 13, 39, 117, 351, ... ________
8. 56, 37, 18, ⁻1, ... ________

Write an equation to represent the function.

9.

x	0	1	2	3	4
y	0	6	12	18	24

____

10.

x	56	57	58	59	60
y	0	1	2	3	4

____

11.

x	0	1	2	3	4
y	15	16	17	18	19

____

12.

x	10	20	30	40	50
y	2	4	6	8	10

____

Draw the next two figures in the pattern.

13. 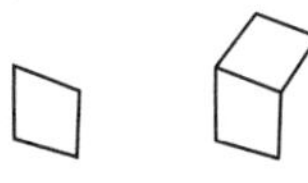

14. 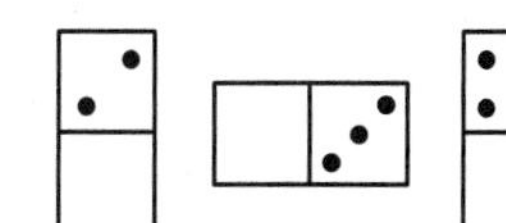

Draw the next two solids in the pattern.

15.

16.

Answers: 1. substract 15; **2.** multiply by 0.6; **3.** add $\frac{1}{3}$ and simplify; **4.** divide by 4; **5.** 53, 60, 67; **6.** $\frac{3}{8}, \frac{1}{4}, \frac{1}{8}$; **7.** 1,053, 3,159, 9,477; **8.** ⁻20, ⁻39, ⁻58; **9.** $y = 6x$; **10.** $y = x - 56$; **11.** $y = x + 15$; **12.** $y = x \div 5$; **13.** ; **14.** ; **15.** ; **16.**

Family Fun PUZZLER

Identify the sequence in Column 1 with its next term in Column 2. Write the corresponding letter on the lines below marked with the exercise number to solve the puzzle.

Column 1	Column 2
1. 3, $^{-}9$, 27, ...	A. $1\frac{2}{5}$
2. $\frac{1}{5}, \frac{3}{5}, 1, \ldots$	B. 9
3. 250, 50, 10, ...	C. 12.2
4. 127, 112, 97, ...	D. $^{-}81$
5. 4.5, 6, 7.5, ...	E. $\frac{16}{81}$
6. $\frac{2}{3}, \frac{4}{9}, \frac{8}{27}, \ldots$	K. 2
7. 4,096; 512; 64, ...	O 8
8. 14.3, 13.6, 12.9, ...	R. 82

___ ___ ___ ___ ___ ___ ___ ___ ___ ___ ___
8 7 1 6 5 4 6 2 3 6 4

Answer: CODEBREAKER

HARCOURT MATH

GRADE 6

Chapter 29

WHAT WE ARE LEARNING

Geometry and Motion

VOCABULARY

Here are the vocabulary words we use in class:

Transformation Movement of a figure that does not change the size or shape of the figure

Translation Movement of a figure along a straight line

Rotation Movement of a figure by turning it around a point

Reflection Movement of a figure by flipping it over a line

Tessellation A repeating arrangement of shapes that completely covers a plane, with no gaps and no overlaps

Line symmetry A figure has line symmetry when a line can divide the figure into 2 congruent parts

Rotational symmetry The property of a figure that can be rotated less than 360° around a central point and still be congruent to the original figure

Name

Date

Dear Family,

Your child is learning to make tessellations, perform transformations of plane and solid figures, and examine symmetry in plane figures.

This is how your child has learned to identify two kinds of symmetry.

What kind of symmetry does this figure have?

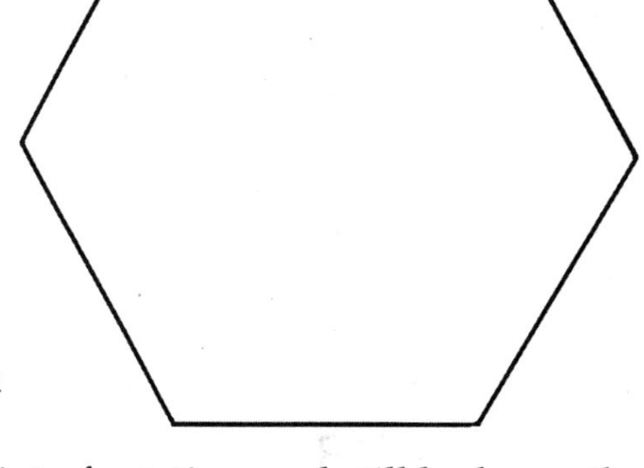

Line symmetry exists when a figure can be folded so that the two parts match, or are congruent.

Rotational symmetry exists when a figure can be rotated less than 360° around a central point, or point of rotation, and still look exactly like the original figure.

This figure has line symmetry and rotational symmetry with a $\frac{1}{6}$ fraction of a turn and a 60° angle of a turn.

Use the model here and the practice activities on the back of this page to help your child practice identifying symmetry. Encourage your child to do this activity with you and other members of your family.

Sincerely,

Identify Symmetry

HOME ACTIVITY

State whether the figure has line symmetry, rotational symmetry, both line and rotational symmetry, or no symmetry. If it has line symmetry, draw the lines of symmetry. If it has rotational symmetry, identify the symmetry as a fraction of a turn and in degrees.

1. 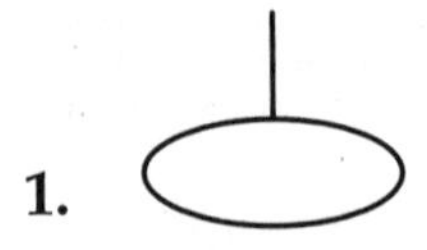

2. 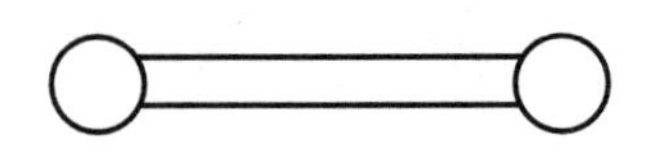

3. 

4. 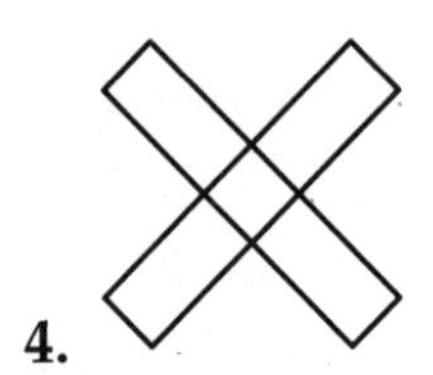

5. 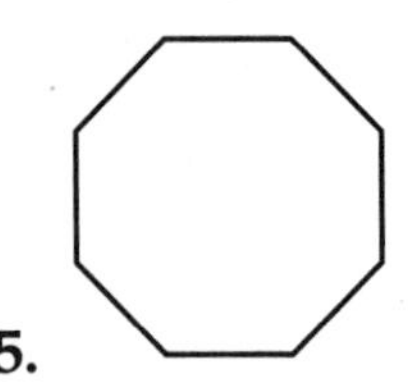

Answers: Check students' lines of symmetry; **1.** line symmetry; **2.** line and rotational symmetry; $\frac{1}{2}$, 180°; **3.** no symmetry; **4.** line and rotational symmetry; $\frac{1}{4}$, 90°; **5.** line and rotational symmetry; $\frac{1}{8}$, 45°

Name ______________________

Geometry and Motion

Tell which type of transformation the second figure is of the first. Write *translation*, *rotation*, or *reflection*.

1. ______________

2. ______________

3. ______________

4. ______________

Tell whether the figure forms a tessellation. Write *yes* or *no*.

5. 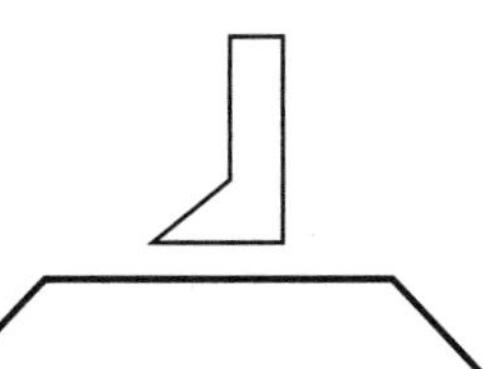 ______

6. 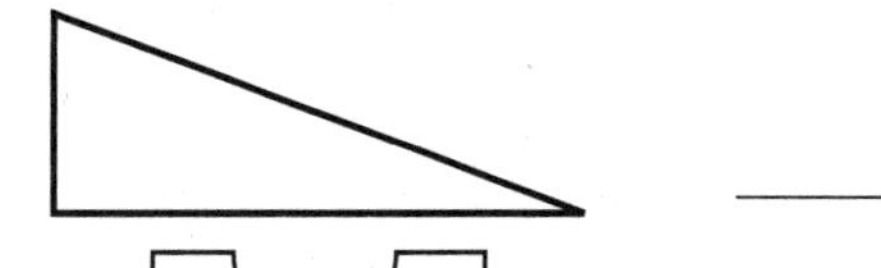 ______

7.  ______

8.  ______

Draw the lines of symmetry.

9. 

10. 

11. 

12.

Tell whether the figure has rotational symmetry. If it does, identify the symmetry as a fraction of a turn and in degrees.

13.  ______________

14.  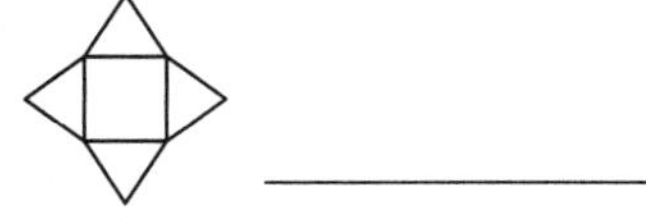 ______________

15.  ______________

16. 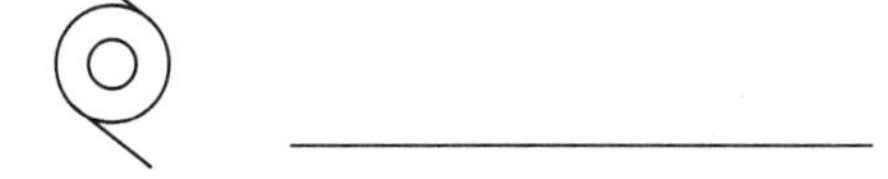 ______________

Answers: 1. translation; **2.** rotation; **3.** reflection or rotation; **4.** rotation; **5.** yes; **6.** yes; **7.** yes; **8.** no;
For **9-12**, check students' lines of symmetry. **13.** yes; $\frac{1}{2}$; 180°; **14.** yes; $\frac{1}{4}$; 90°; **15.** no; **16.** yes; $\frac{1}{2}$; 180°

Family Fun The Name Game

Reflections	Print your name horizontally. Make a horizontal reflection of your name. Print your name vertically. Make a vertical reflection of your name.
Rotations	Print your name horizontally. Make a 90°, 180°, and 270° rotation of your name.
Tessellations	Identify any letters in your name that might tessellate when you use block letters. Draw the tessellation.

HARCOURT MATH

GRADE 6

Chapter 30

WHAT WE ARE LEARNING

Graphing Relationships

VOCABULARY

Here are the vocabulary words we use in class:

Inequality A mathematical sentence containing <, >, or ≠ to show that two expressions do not represent the same quantity.

Coordinate plane A plane divided by a horizontal *x*-axis and a vertical *y*-axis.

Ordered pair A pair of numbers used to locate a point on a coordinate plane. The first number tells you how far to move right or left from the origin. The second number tells you how far to move up or down.

Axes The two number lines that divide a plane into quadrants

Quadrants The four sections of a coordinate plane

Name

Date

Dear Family,

Your child is learning about graphs. Your child will solve algebraic inequalities and use ordered pairs to describe locations on the coordinate plane and to show relations. We will explore linear and nonlinear relationships and graph transformations.

Here is how we graph relations:

Complete the function table showing a relation. Then graph the data on a coordinate plane and write an equation relating y to x.

x	2	1	0	-1	-2
y	10	5	0	?	?

- **STEP 1** Identify the sequence. Complete the table.
- **STEP 2** Write the data in the table as ordered pairs. (2,10), (1,5), (0,0), (⁻1, ⁻5), (⁻2, ⁻10)
- **STEP 3** Plot the points on the graph.
- **STEP 4** Write an equation relating *y* to *x*. $y = 5x$

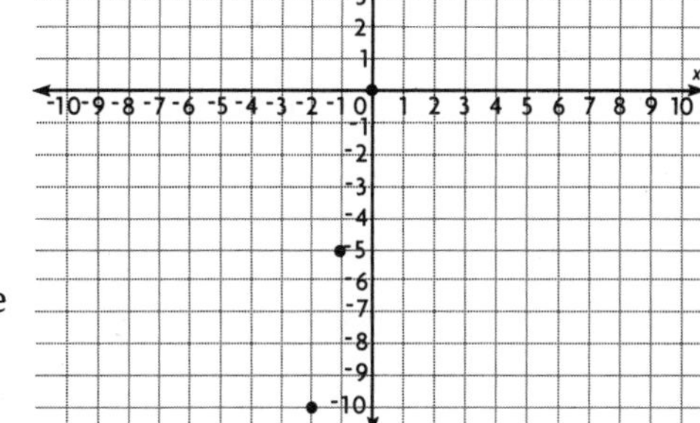

Use the model here and the activity that follows to help your child practice graphing relations. Encourage your child to do this activity with you and other members of your family.

Sincerely,

Seeing Relations

HOME ACTIVITY

For each problem, complete the function table. Then graph the data on a coordinate plane and write an equation relating y to x.

1.

x	0	1	2	3	4
y	0	4	8	12	

2.

x	0	$^-2$	$^-4$	$^-6$	$^-8$
y	8	6	4		

3.

x	0	2	4	6	8
y	0	$^-2$	$^-4$		

4.

x	2	1	0	$^-1$	$^-2$
y	$^-2$	$^-3$	$^-4$		

5.

x	10	8	6	4	2
y	5	4	3		

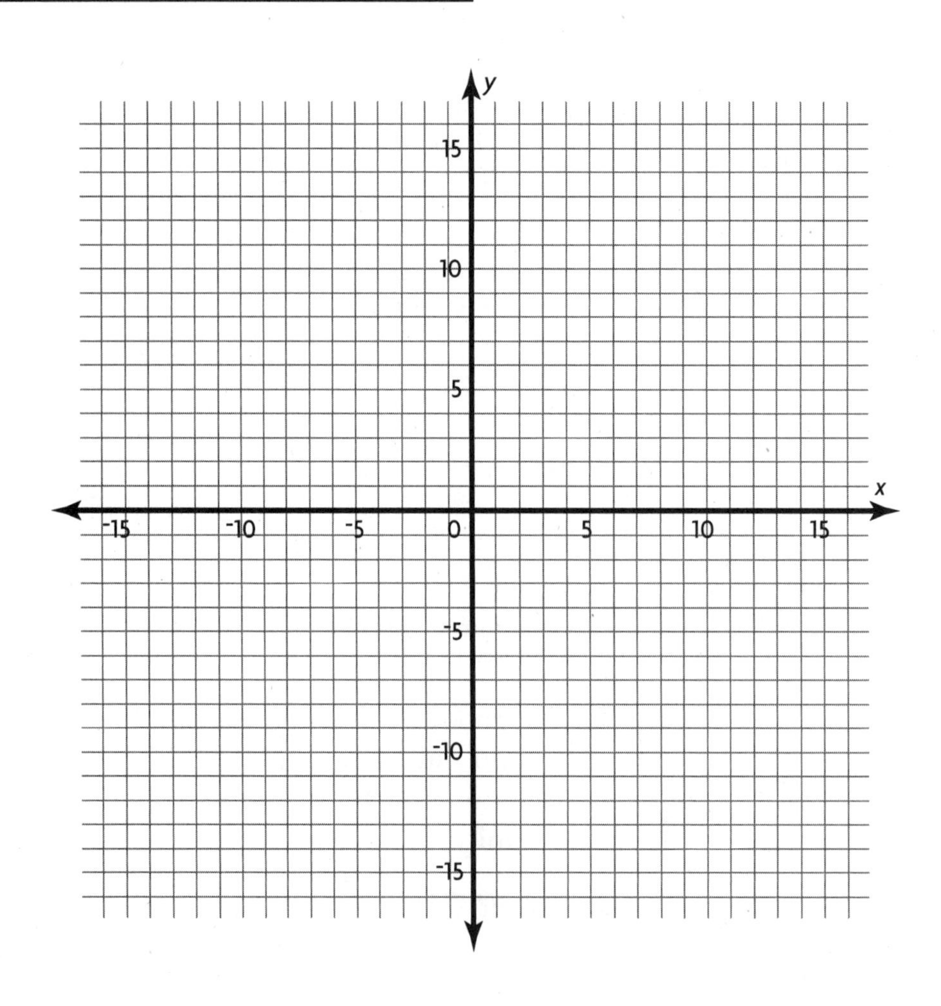

Answers: For 1-5, check students' graphs. **1.** $y = 4x$; **2.** $y = x + 8$; **3.** $y = {}^-1x$; **4.** $y = x - 4$; **5.** $y = \frac{1}{2}x$

Name ______________________ PRACTICE/HOMEWORK

Graph Relationships

Graph the solutions of the inequality on a number line.

1. $x > 5$
2. $x < 3$
3. $x < 6$
4. $x > 2$

Solve the inequality and graph the solutions on a number line.

5. $t + 3 < 4$ ____
6. $\frac{m}{2} > 4$ ____
7. $p - 6 > 0$ ____
8. $3d < 15$ ____

Describe how to locate the point for the ordered pair on the coordinate plane.

9. $(6, 3)$ ________
10. $(^-3, 4)$ ________
11. $(5, ^-2)$ ________
12. $(^-3, ^-3)$ ________

Write the ordered pair for each point on the coordinate plane.

13. point A ________
14. point B ________
15. point C ________
16. point D ________
17. point E ________
18. point F ________
19. point G ________
20. point H ________

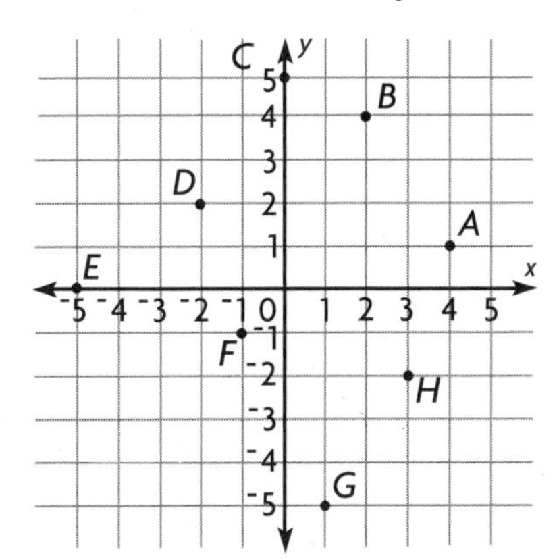

21. Complete the table.

x	1	2	3	4	5
y	5	6	7		

22. Write an equation that relates y to x. ________

Solve.

23. Triangle ABC has coordinates A (2, 3), B (2, 1), and C (4, 1). What are the new coordinates after it is translated 2 units right and 2 units up?

24. Trapezoid $DEFG$ has coordinates D $(^-2, 2)$, E $(^-1, 2)$, F (0, 0),and G $(^-3, 0)$. What are the new coordinates after is it reflected across the y-axis?

Answers: For **1-8**, check students' graphs. **5.** $t < 1$; **6.** $m > 8$; **7.** $p > 6$; **8.** $d < 5$; **9.** right 6 and up 3; **10.** left 3 and up 4; **11.** right 5 and down 2; **12.** left 3 and down 3; **13.** (4, 1); **14.** (2, 4); **15.** (0, 5); **16.** $(^-2, 2)$; **17.** $(^-5, 0)$; **18.** $(^-1, ^-1)$; **19.** $(1, ^-5)$; **20.** $(3, ^-2)$; **21.** 8, 9; **22.** $y = x + 4$; **23.** A' (4, 5) B' (4, 3), C' (6, 3); **24.** D' (2, 2), E' (1, 2), F' (0, 0), G' (3, 0)

Family Fun

WHAT IS IT?

MATH GAME

Plot points for the ordered pairs in the order given, working left to right. Connect the points to reveal the mystery object.

(⁻5, 10)	(⁻2, 9)	(0, 8)	(2, 7)	(4, 5)
(5, 3)	(5, 0)	(5, ⁻3)	(3, ⁻5)	(1, ⁻7)
(⁻2, ⁻9)	(⁻2, ⁻12)	(⁻2, ⁻15)	(⁻3, ⁻13)	(⁻5, ⁻12)
(⁻7, ⁻11)	(⁻8, ⁻9)	(⁻6, ⁻7)	(⁻4, ⁻6)	(⁻5, ⁻3)
(⁻6, 0)	(⁻5, 3)	(⁻4, 5)	(⁻3, 7)	(⁻3, 8)
(⁻5, 10)				

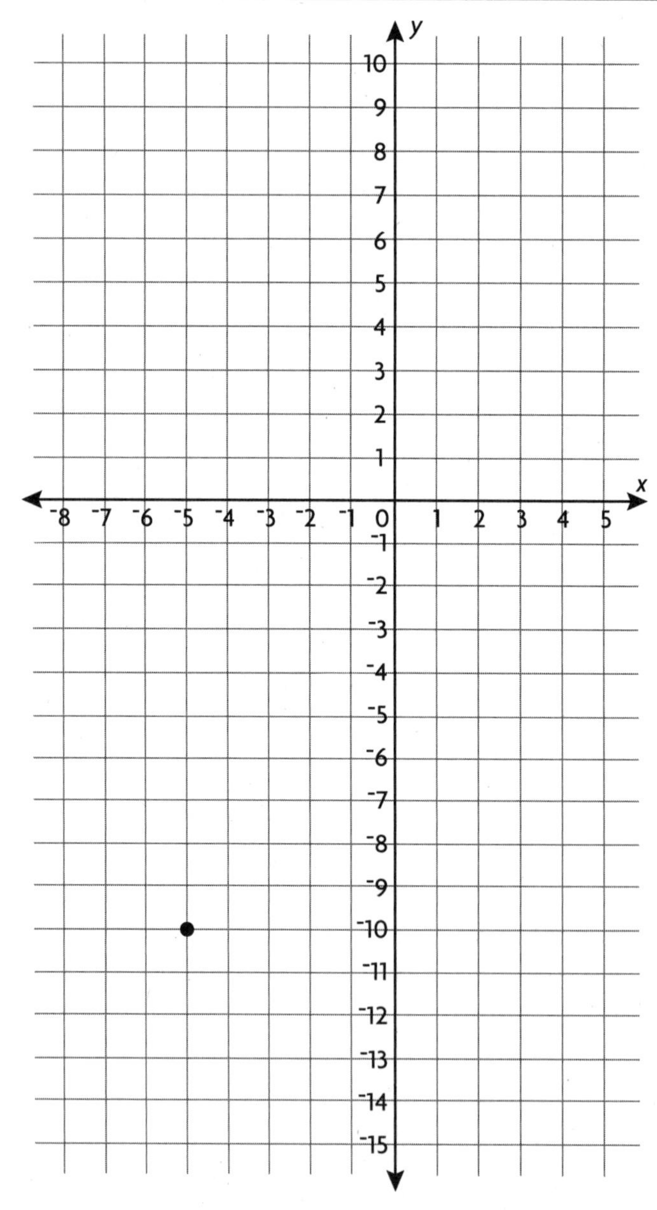

Answer: a duck